THE 5-STAR BUSINESS NETWORK

Vivek Sood

PRAISE FOR THIS BOOK:

The speed of innovation is multiplied many times over - the head of Airbus recently complained that new Airbus planes coming into the market are already outdated in terms of ICT (information and communication technology). Use of 5-STAR Business Networks help to multiply this speed of innovation and if your company is not using these 5-STAR Business Networks for Innovation then it is likely very soon your company will fall behind your competitors. That is how the chemicals companies are changing their business models from sellers of commodity chemicals to solution sellers, and most other industries are creating new business models today.

Dr Hanns Zeltinger
Former CIO DaimlerChrysler,
Financial Services Asia Singapore

The rules of the business world have changed - and few have a grasp of the new reality. This book is a pivotal wake-up call for business leaders to get them on track towards a whole new step change or two in performance. From the use of technology to outsourcing to innovation - the breadth and depth of thinking is seminal and intense. The best aspect of this book is the actionable nature and practicality of the content.

Tony Fedorowicz
Ex Head of Supply Chain Asia/Australia -
Orica Mining Services, Sydney

In the high volatility environment, if you are running a portfolio of supply chain purchases worth hundreds of millions of dollars a year, there is no other way to mitigate the enormous risk than by managing the business network extremely carefully. This book expounds and builds on some of the techniques I have utilised for risk management and value creation in a soft commodity supply chain through roller-coaster highs and lows of the recent commodity cycles. There is no doubt Business Networks are powerful beyond belief and useful beyond comparison.

Hugh Robertson,
Ex Director - Commodities and Risk, Weston Milling, Sydney

This book offers a breath of fresh air in the world of business strategy dominated by the staid Ivy League thinking which cannot see beyond its very own doorsteps. When everybody is thinking alike - in general - nobody is thinking at all. This book will force you to think afresh, and it will show you how to apply your thinking in the changing business environment to create new business models that suit the modern business reality a lot more.

PRAISE FOR THIS BOOK

Gennady Teplitsky
Ex Vice President - JPMorganChase New York

From the outset, it is clear that Vivek Sood brings passion, intelligence, and uncommon insight to his topic - understanding and interpreting for the reader the powerful influence and value of the business networks that underlie success in the modern world. Well-researched, broad-reaching, and with a refreshing "outsider" perspective free of academic jargon, this practical book will give you insights you can use to create and leverage these powerful networks in your own business - a must-read for anyone who wants to understand and use one of the most powerful forces at work in today's hyper-connected world.

Tony Coretto
Co-Founder and Co-CEO, PNT Marketing Services

I have been most impressed with the business acumen and insights in these pages. The author has clearly become an authority in his logistics and management field worldwide. His book is a gem to be read and appreciated by all those who desire to be on top of current managerial thinking and desire to reap the benefits of his experience and capacity. I recommend this to all. I certainly will use this book in my practice and benefit from it personally. I recommend it to my students interested in consulting, operations and logistics. A must and worthwhile read!

Lin Giralt
Adjunct Professor and Lecturer, Rice University
Jones Graduate School of Business, Houston

This book shines out of the pile not because it is an in-depth and original look into the power of the business networks, but because its author applied in it his rare passion, vision, common sense and innovation, which are core for today's business evolution, achieving results and leading and inspiring for a better world. Read this book if you want to be inspired in your daily activities, regardless of whether you are a chairman or a student - it has it all for everybody!

Andrey Manev
CEO, HUS

5-Star Business Networks is the new model that makes the existing model obsolete. In these rapidly changing times, new systems and tools are needed. Authentic leadership and emotional intelligence are being used within global 5-Star Business Networks to build trust and better results. This is the new economic currency.

Phil Johnson
Organisational Development Author, Speaker, Teacher and Coach President, Master of Business Leadership Inc., Toronto

Author Profile
VIVEK SOOD

For over 25 years Vivek Sood has empowered corporations on 5 continents to achieve their peak potential – in operations, strategy, and all components of their entire supply chains. In January 2000, he co-founded and managed Global Supply Chain Group, a high impact services company made up of supply chain pioneers and thought leaders who work only on selected high impact strategy projects with some of the largest corporations in the world.

Prior to that, Sood was a management consultant with top-tier strategy consulting firm Booz Allen & Hamilton. The Sydney resident spends 60% of his time travelling around the world with a single passion – to create, configure, and formulate effective, secure and sustainable supply chains around the world. A dynamic, high impact professional and presenter, Sood has a unique ability to shake-up the status quo and transform operations to maximise returns. His work experience spans more than 80 countries on 5 continents, and the clients range from fortune 500 companies to some of the best known brands on earth.

He has also worked with some of the most innovative green technology companies to help create outstanding supply chains from scratch for businesses worth hundreds of millions of dollars, under tight time constraints. His work has added cumulative value in excess of $500M, incorporating projects in major supply chain infrastructure investment decisions, profitable growth driven by global supply chain realignment, supply chain systems, negotiations and all other aspects of global supply chains. Sood is also an acclaimed author of two very highly regarded books. He writes regular path breaking articles and lectures at business schools, supply chain conferences, forums and workshops in various parts of the world.

SPECIALTIES

Business and Supply Chain Transformation -
helphing companies multiply profits using the full power of global business networks.

"Vivek's company, Global Supply Chain Group, created outstanding results in configuring global supply chain where none existed in an industry which is still finding its moorings. While the investments are huge, the responsibility to ensure these are wisely made is even bigger. Thanks to Vivek's team, and its culture of working seamlessly with internal leaders, we successfully ensured our investments were wisely made."

Bernhard Voll, *Director of Engineering, Graphite Energy*

"Vivek is one of those rare top-tier thought leaders in the world of Supply Chains who combine immense practical experience with a very high level strategic thinking to create unique and invaluable supply chain solutions. The fact that he does it in such an easy going, friendly manner is an added plus."

Ben Gordon, *Managing Director, BG Strategic Advisors*

[More details of Vivek's career are available on line on http://www.linkedin.com/in/vivek]

THE 5-STAR BUSINESS NETWORK

www.5starbusinessnetwork.com
Published by: Jardine Thompson Pty Ltd,
Level 9, 80 Arthur Street, North Sydney, NSW, 2060, Australia

ISBN-13: 978-0615794198
ISBN-10: 061579419X

Typesetting by: Anaïs Lelong
Cover Design by: Thomas Radrigan

MOVE BEYOND TRADITIONAL SUPPLY CHAINS

The 5-STAR Business Network

And The CEOs Who Are Building The Next Generation of Super-Corporations With It

by Vivek Sood

Jardine Thompson

DEDICATION

This book is dedicated to the future generations - they will make the world an even better place. In particular this book is dedicated to our sons - Raman, Jay and Dev - who make every day for me a joy to live.

"Life is no brief candle to me.
It is a sort of splendid torch which I have got a hold of for the moment, and I want to make it burn as brightly as possible before handing it on to future generations."

George Bernard Shaw

CONTENTS

Introduction .... **XXXI**

SECTION 1 - NETWORKS **1**

Chapter 1 The Most Important Network on Earth 7

Chapter 2 Open Secrets from Ancient Networks 21

Chapter 3 Lessons from Modern Networks 33

Chapter 4 The Hidden Power of Business Networks 45

Chapter 5 Truth about the IT Systems Required for Great Business Networks 65

SECTION 2 - 5-STAR BUSINESS NETWORKS **99**

Chapter 6 What Super Networked Businesses of Today Can Teach 109

Chapter 7 Five Key Cornerstones of The 5-STAR Business Network 125

Chapter 8 Why Innovation is Not a Solo Sport 147

Chapter 9 Cash Rules 177

Chapter 10 Co-Multiply Your Profits - Use the Full Power of the Global Business Networks 197

Chapter 11 Do Not Be a One Hit Wonder 215

Chapter 12 Outsmart, Outsource, Outprofit 237

Section 3 - Practical Example of 5-STAR Business Networks 267

Chapter 13 Super Networked Business of the Third Richest Man on Earth 273

Chapter 14 Research Methodology and Other Companies With 5-STAR Business Networks 315

Appendix 325

Notes 333

Acknowledgement 345

Index 351

"I love QUOTATIONS because it is a joy to find thoughts one might have, beautifully expressed with much authority by someone recognised wiser than oneself."

Marlene Dietrich (1901 - 1992)

THE 5-STAR
BUSINESS NETWORK

A NOTE TO THE READER

Dear reader

Congratulations. By taking this book into your hands, you will discover many new concepts applicable in the real business world. It is not written for everyone, but it is an absolute must to own and read for those who want to embrace the developments of traditional businesses into modern networks, to go beyond existing boundaries by adopting the key principles of modern supply chain management and to understand why being a stable and trustful part of a network can multiply the value of your own business many times over.

Business networks without legal or political restrictions will dominate and determine our future. The needs of individuals and the ultimate end-users of the value chain will be in the foreground and shift the priorities of the past, where manufacturers and suppliers of products dominated the business. Today's business models must be turned upside down to create new and effective structures, including innovative procedures. Supply Chain Management and Supply Chain Excellence, as well as innovative Network Management, are the keys to success.

This book will show you where and how to walk into the future with your business. Proven and good examples lead the way forward, but do not copy them, take your own route or take a scout to avoid mistakes and disappointments. If you are on the right track,

I can guarantee you to multiply the value of your business by many times. Being a trustful partner in a stable and reliable network increases the value of a company to unbelievable dimensions. Investors will pay you a premium for this strategy! And the costs will decrease dramatically because you use the strengths and resources of other network partners without investing in expensive capacities.

Therefore, this book can change the world into a better one if the messages are understood and applied thoroughly.
Be excited and enjoy reading!

Dr Wolfgang Partsch
Co-Founder of Supply Chain Management

PREFACE TO THE SECOND PRINT

The response to the first print has been overwhelming, which has made this second print necessary in a short period of time. I am thankful to all the managers, executives and CEOs who have taken time to comment, recommend and pass on this book to their colleagues. Primarily, they are the target audience of this book and their engagement has been extraordinary as seen from the numerous emails and positive comments that I have received for this book, as well as the brisk dispersion of this material within this group. This certainly makes the effort to write, revise, edit, proof and revise this book worthwhile. Many of the readers have commented on extraordinary pressures faced by the executives and CEOs, and why new business models discussed in this book are so important to them.

Executives, CEOs and boards carry an extraordinary amount of burden in today's society. There is no doubt that they are the primary producers of wealth in the society, via the companies they lead and products those companies create. Not many other forms of organisations create or commercialise products and services to any great extent. Yet, the CEOs carry a huge burden.

First, there is pressure to do more with less. The constant market

Not just the boards, but also other key shareholders and pressure groups can all create an environment where CEOs and key executives simply must produce more results with less and less resources. New technologies and techniques, new way of doing things, and new business models are the only savior in this environment. While most executives adopt this change with the proper diligence, rarely do these new business models and techniques emerge from the Ivy league, or other traditional sources of knowledge. Most CEOs have to search around and find their own sources for these new business models and techniques.

Second, as the traditional command and control organisations of the past have broken down, leadership of generation "Why" has become an intense challenge. With the current eroding organisational loyalty, a culture of job-hopping and attitude of cynicism, leadership challenges facing today's CEOs are far more intense than faced by any generation in the past. While the socio-cultural causes and effects of these changes are beyond the scope of this book, the challenges they bring to today's CEOs are worth noting here because of their impact on the growing pressure on CEOs' top executives.

Third, the intense scrutiny from the pressure groups, NGOs, environmental lobbyists and other such sundry groups has created an atmosphere where the "business of business" is becoming increasingly difficult. While serving a useful function in many instances, it is easily possible to go overboard with all this activity so that the playing field is tilted in favour of businesses in countries where such activity is not permitted or possible. The CEOs and executives have to live with - and work amongst - all this intense activity and operate successful businesses under the very vocal scrutiny of self-appointed guardians of societal wellbeing.

Finally, the CEOs face a unique situation where the buck stops with them, but they rarely get told the pure, unadulterated truth. I do not want to say that everyone lies to his boss, but the reality is that in most instances, almost all the internal and external messengers know that their own career progression and economic well-being the helm. Under such circumstances, it is quite easy to temper the message and carry only the good news, lest the message harm the messenger. Most companies survive a long time while operating in this mode. However, if a company fails to face unpleasant reality in good time, it is quite easy to fall behind in the marketplace and gradually lose privileged market positions.

When supply chain management became popular, it was seen as a new way of doing business because it helped break down the barriers between various silos within the business. The extraordinary success of the discipline of supply chain management is attested by the fact that today nearly 87 percent of the business schools, even the ivy league institutions, offer a course in supply chain management or one of its related disciplines.

But, the most progressive executives and CEOs are looking beyond breaking the internal barriers now. They are seeking to work seamlessly with teams of internal and external personnel in order to rapidly create products and services, and then profitably deliver them to appreciative customers. The teams could come from multiple companies – large or small – and work together for a short or a long time for their intended purpose.

The feedback to this book has consistently emphasized this aspect of the content, where readers have created new business models and used them to increase profitability, turnaround stressful situations, and salvaged businesses. While most consultants, especially from larger top-tier companies, are reluctant to put their best ideas out because they like to make once and sell multiple times, I face no such pressure. Firstly, my own time is limited and there is no way I can work on all the projects that can use my ideas expressed in this book and in my other books. Secondly, the CEOs who use these ideas can still find a way of harnessing them with my assistance, if they choose to do so. Finally, an idea is only as good as it proves to be in practice. After all, when Booz Allen and

Hamilton created the original idea of supply chain management they put it in the public domain leading to a vast revolution in how companies do business today.

This is my hope and desire – that you will use what is usable for you in this book, and share with me how you used it. This is also important because we are all learning from one another and constantly improving this material, and noting what other users can learn from your experiences. To share, to comment, to commend, to discuss, or to ask, send an email to me at v.sood@globalscgroup.com. While you already know this, in today's world there is no such thing as one-way communication. That has been my primary learning from the feedback I got so far from this book. I am looking forward to more of the same.

Some executives who received a copy of the book have enquired how they can contribute. In my view, there is no better way than paying forward - share the ideas, the content, web-links, even the book with your own business network of executives who can benefit from these concepts. There is no doubt that on this round planet, what goes around comes around. You will not only enrich your own business network, but also yourself in this process. After all, in this modern inter-connected economy, your network is your net worth.

To your continued success,

Vivek Sood
Sydney, Australia
October 14th, 2013

PREFACE

> ***I have no special talent.***
> ***I am only passionately curious.***
>
> Albert Einstein

Most effective business leaders relish the challenge of answering questions such as the following:

- Why did Apple sue Samsung while it continues to buy critical parts for its winning products from Samsung?
- Why did Google create Android OS for mobile applications, and is now talking about opening its own retail stores?
- Why did Amazon create Kindle when the market was already saturated with other tablets and similar products?
- How did Nokia Mobile phone lose its shine?
- Why did Apple build its own retail presence?

- How will shale gas discoveries in North America change the business world and perhaps the geopolitical balance in the next 10 years?

If you are as deeply passionate about the world of business and supply chain networks as I am, and enjoy exploring similar questions, and coming up with answers that will help immensely in using this wisdom to build your business, then join me. This book is a journey of exploration through the world of business networks that run the commercial world today.

When General Motors filed for Chapter XI protection in 2008, it also marked the closing of a chapter in modern commerce. General Motors was seen as the paragon of modern American management theory as popularised by Peter Drucker in the middle of the 20th century. It was at this venerable company that Drucker formed his early thoughts about management as a profession, separation of the ownership from management of enterprise, the key functions of management, division of labour, theory of leadership of enterprise, indeed the very concept of the corporation. His writings filled the needs of the time, and were picked up by the Ivy League business schools and corporations alike, to form the basic foundation of management profession.

Indeed there was a time when General Motors and US commerce were thought of as interchangeable entities, with the popular aphorism that "what is good for GM is good for America and vice versa." Some people still think this is the case. They see the decline of General Motors as symptomatic of a wider malaise in the US economy. Others think General

Motors will rise like a phoenix to become an industrial powerhouse.

While we do not know what will eventually happen to General Motors, we know that new models of commerce, new industries, new technologies and new ways of solving old problems will be required to build a stronger economy at a global level. All of these will not necessarily come out of one country, one continent or even one region.

Drucker foresaw some of these changes, as shown in his writings on the information age, post-capitalist society and post-industrial man. Prescient as he was, he did not yet fully see the majority of the changes that have happened since his death in 2005. The rise of China and India, the global financial crisis, the zombification of the Western economies as a result of the intense focus on the rapid gains from the FIRE (finance, insurance, real estate) industries, hollowing out of real capabilities are nowhere to be seen in his writings. However, this is not just true of Drucker; a majority of the management thinkers, writers, academics and authors can be painted with the same brush.

It is not a surprise that the established minds find it difficult to think outside the box. Since the times of Aristotle, Socrates and perhaps even before that, new ideas must come from new places - from outside the established order of thinking. No wonder then, that the most innovative companies in the US still choose to locate on the West Coast. Many of the most successful corporations were formed by college drop-outs and the most successful business models do not even have names yet.

In this book, we will not only name some of these models,

but we also will study them closely and understand these new methods. The aim is not to copy or even emulate, but to gain a deeper understanding of their workings, to learn how to formulate similar models under different circumstances that your company faces.

At this point, you may ask, 'What entitles you to write this book?' Let me include a couple of brief biographical paragraphs that will hopefully spur you on to devote your full attention to get the best use out of this book.

The foregoing paragraphs might have already signalled to you that I see myself as an outsider. I do not have a Bachelor's degree and did not really start my college education until the age of 29 - but both circumstances helped me succeed immensely in life. At the age of 17 and straight out of a boarding school in Northern India, I joined a merchant marine company as a deck cadet after a brief training.

My first assignment was to go to a shipyard in Japan to work with the ship's captain, chief engineer and other officers to take delivery of a brand-new vessel being built there. At this time, due to the severe depression in the global shipping industry, I saw many super tankers worth million dollars sailing on their maiden voyage straight from the shipyard to the demolition yard.

It was like seeing a baby being killed at birth. It aroused in me an intense desire to study the causes of economic cycles - I had already seen their effects. Over the next three years, I circumnavigated the globe many times on ships, working 12-16 hours a day 6-7 days a week. I spent all my spare time reading voraciously about economics, finance and modern commerce. I kept my day job for another eight years, as it afforded me the luxury of visiting exotic ports and harbours, where I had experience working with stevedores, logistics companies, commodity producers and traders, importers, exporters and various other wheelers and dealers around the world. I earned a good living - enough to put some money aside for what I then considered a proper education.

In 11 years I travelled to more than 100 countries, rose in rank from a cadet to a master mariner with progressively more responsibility and a wider perspective. I saw thousands of different ways to do similar things in the different places I visited and worked. All of these methods had some qualities worth emulating, yet there were many others that were of little value. In my spare time, I added another pursuit to my economic studies – combining alternate ways of doing things to find the best method to suit the circumstance in question. These two pursuits had already become my twin passions long before I studied for my MBA at one of the best business schools in Australia and took a job as a management consultant in Sydney with a top-tier global consulting company.

For the past 15 years, I have been a management consultant to CEOs, boards and top executives in Asia, Australia and other parts of the world. I have focussed on empowering corporations on five continents to achieve their peak potential – in operations, strategy and all components of their entire supply chains. In January 2000, I co-founded and managed Global Supply Chain Group, a high-impact services company made up of supply chain pioneers and thought leaders who work only on selected high-impact strategy projects with some of the largest corporations in the world. I spend 60 percent of my time travelling the world for a single passion – creating effective, secure and sustainable supply chains.

With that said, let me also use one paragraph to talk about why I wrote this book. With more than 10,000 business books released each year, it is clearly not a worthwhile exercise to add to the cacophony of sound, unless you have a meaningful message to convey. The returns are paltry and I have already written another book, listed with a respected publisher under my name (albeit with shared credits as a co-author). I do not particularly relish the long work hours any more. What motivated me to spend nearly every spare moment in the past 18 months to write this book?

Business is the engine of society and it will need newer models of commerce to fast-track any recovery. The hollowing out of skills from entire societies, without replacing them by other competencies, high levels of youth unemployment (often disguised by serious-looking play on iPads and tablets), and growing economic imbalances risking implosions of unrest, civil commotion or even a great war, have all combined to create an alarming set of circumstances. Whatever transpires in the short-term, eventually business people will have to lead the way to recovery around the world.

It is evident that I am passionate about newer models that work better. In my projects and work around the world, I have noticed that, in almost all circumstances, there is always some way to make things better. We only have to look around and see where the guidelines and trends are and which models will best suit.

Eleanor Roosevelt (1884-1962) famously said:

" ***Great minds discuss ideas,***
Average minds discuss events,
Small minds discuss people. "

While it is very tempting to discuss just one of the three key ingredients of life - ideas, events, or people - in this book we discuss all three because they are inextricably linked.

People make events and create ideas. Ideas create events and help people become successful, events shape people and give impetus to ideas. A book full of concepts and ideas, with no stories about people or events, would be extremely boring and dry. On the other hand, a book with just chronicles of events or people would hardly be worth bothering to read unless its author had an immensely entertaining style of writing (which I do not), and even then would be of little practical value besides entertainment. I mostly use events and people to illustrate ideas and concepts to make them more tangible for readers.

Primarily, then, this book is about ideas and concepts - yet you will see enough discussion about people and events to be able to use the concepts. Most of the people and events discussed are relatively well-known, so that background contextual information is already present in the readers' domain. Occasionally I use events from case studies based on our work - only because we could not find a well-known event illustrating the concept. I do not make apologies for that, or for disguising some data or names of the entities for obvious reasons of confidentiality.

Every couple of decades, powerful juxtaposition of trends leads to unique and revolutionary way of commerce. Contrary to the gushing accounts and adulations of the commentators, most pioneers merely stumble on these trends by a process of trial and error. Other companies, the more nimble and hungry ones, follow the pioneers closely and build strong businesses in their lead. More established companies then follow suit and try and recover lost ground, using their financial muscle and market power, sometimes succeeding and sometimes failing. Many other companies are so caught up in the hubris of their past success, internal politics or other attention-sapping device they fail to move at all, or move too late, often with disastrous consequences.

Chronicles of such disasters would perhaps be more instructive than the starry-eyed accounts of success. As Daniel Coyle points out in his book, *The Talent Code*, the only way to succeed massively is by failing repeatedly at progressively more complex, smaller tasks until you master them. However, most writers and authors persist with the formula that has succeeded since the first bard told stories of the victories of war, and I have no doubt that we will continue to see many romantic accounts of success for centuries to come.

It is difficult not to get caught up in the current of adulation that surrounds some companies at certain times. Any such apparent adulation in this book is despite my effort to be objective and cognisant of the cyclical nature of success.

As defined earlier, the aim of this book is rather more solemn. It is to take a wide and deep perspective on business trends, define the useful trends as seen from user perspectives, and come up with useful information for the business executives and managers.

The biggest trend sweeping the world - business and non-business - today is networking. When Facebook announced its IPO filing, it valued the company at $100 billion. Analysts, pundits and business school professors are still debating what entitles it to that kind of valuation when many web-sites with similar business model - Orkut, Myspace and a plethora of wannabes - failed to monetise the eyeballs to any great extent. It was not even validated how many of those numerous Facebook accounts were authentic.

Other contenders to high valuation - LinkedIn, Twitter and a few others - do not command the stratospheric valuation claimed by Facebook. LinkedIn was valued at "mere" $9 billion, while Twitter languishes in third place among the winning triad at a "meagre" $8 billion.

These high valuations, whether justified or not, inspire awe and wonder. How can a company which has no manufacturing, no traditional customers, no suppliers, no products, no warehouses and no logistics, command a valuation roughly equivalent to 10 percent of the GDP of the entire continent of Australia? By what magic is the projected cash flow so high, or, the risk-adjusted cost of capital so low? How can a company that did not even exist 10 years ago be suddenly more valuable than roughly 95 percent of the corporations in the USA - many of them earning billions of dollars of steady cash every year for decades, if not centuries?

Our purpose, however, is not to justify or critique the valuation of these social networking enterprises. It is to look at the power of the networks in the business world. Here, we will discover a set of Business Networks even more valuable than those mentioned above. In fact these Business Networks, rarely recognised at first, are so valuable that not a business can exist without one in today's world. Unfortunately, most executives and CEOs fail to harness the full potential of this network adequately.

This book is about the 5-STAR Business Networks that run the modern world. Is your company reaping the full rewards of all the effort put into the activities within the company? Are you happy with the results - sales, profits, costs, competitive positioning, new product pipelines, supplier performance and other myriad details of your company's business? This book is for every executive, every CEO, every manager and every management student - no matter which business they are working in, or which part of the business they are part. I can scarcely think of a business situation in which the concepts of this book will not lead to a massive, positive shake-up of the status quo. After all, progress is all about discarding the tired, old and obsolete ways of thinking and doing things.

INTRODUCTION

When Vikram Pandit, the former CEO of $1.84 trillion Citigroup announced his decision to leave the bank, I was told a startling statistic by a colleague - the average time between jobs for CEOs is over 19 months. A similar, though smaller number applies to other C-level executives. The reasons for this range from gardening leave and non-compete clauses, to lack of suitable opportunities at the same level. Executive placement firms and business schools continue to research the topic of CEO/executive tenure, succession planning, career planning and movement which is always of intense interest to the executives themselves. However, there is another interesting side-effect of this trend.

Within their extraordinary network, many highly capable executives find ways of adding immense value to the corporate world as consultants and advisors. Many C-level executives who work alongside our teams on strategy projects have enjoyed the sojourn free of corporate politics and focussed intently on value creation for the clients. Most of them are independently wealthy and want to contribute where their talent can best be used for maximum value creation. Others are fed up with shenanigans and Machiavellian manoeuvering and want a temporary reprieve. In our company, we welcome them with open arms. They enrich us with their knowledge, expertise and contribution to our clients' businesses.

A lot of material in this book comes from what we have learned and applied in business and our clients' businesses in these networks. The trend towards a freelance economy is just starting to make news. Even executives with decades of loyalty to a single company do not think in terms of jobs any more - much of the work in corporations is now oriented around projects. The teams of people working on a project within a corporation can come from a multitude of companies, or from a growing pool of freelancers.

Business schools, economists, think tanks and researchers are starting to take notice and will, no doubt, eventually produce reams of paper covering the freelance economy from all angles. A lot of concepts and their applications covered in this book are very new and will stimulate the research referred to above. I make no apologies for the relative newness of the material and for not having hundreds of case examples backing up every conclusion. By the time that kind of research would be conducted, trends will have moved further and the conclusions will be merely of historic interest to most executives who live their lives on the leading edge and are the intended audience of this book. If some of the concepts of this book stimulate academic research, I do ask that they be properly acknowledged.

So what can you expect from this book? You probably already have a network of hundreds, if not thousands, of business associates and wonder what more can you do with networking. In the new world, it is not enough just to know people.

They are not going to call you when they need you. Let me illustrate this with a personal story. A client who was a C-level executive was in tricky situation. The large corporation he worked for had chartered a ship to carry its cargo. As a result of an accident, the ship was crippled. This lead to a massive loss of property and some injuries. The company went through a testing time to make good and clear up the mess, deploying specialist services including marine surveyors, loss adjusters, divers, shipping agents, port captains, arbitrators and lawyers. The total bill ran into tens of millions of dollars. When the same executive later told me what had happened, I was flabbergasted. I asked him: "You knew very well my extensive and intensive shipping industry background. Why did you not call me, I could have saved your company a lot of money". His reply was enlightening. After some reflection he said, "Vivek, I never thought of that! If I had, I would most certainly have called you, and yes, in this situation we would have saved a lot of money, if I had made that phone call."

My insight from this conversation is that a network is not sufficient unless your value-added potential is transparent and has top-of-mind awareness for your network. This book shows you a multitude of ways in which business networks add and communicate value in the modern business world.

In particular, I will show the five useful business trends, and new ways of looking at them, to create, grow and reap outstanding and sustainable business advantages.

The first business book I read was "*In Search of Excellence*" by Tom Peters and Robert Waterman. It was a gushing account by two ex-McKinsey consultants, truly in search of excellence among American businesses, and contained a plethora of advice that to my then untrained mind (after all, I was just an untutored merchant navy officer) appeared rather obvious - for example, walk around your operations to see what is going on.

What struck me most about the book was that in the intervening 13 years or so between the time it was written and the time I read it, most companies singled out as excellent by the authors were already in trouble. That impression - that companies once lauded as excellent can rapidly lose that mantle - has never left my mind. I have read more than 5000 business books, countless book summaries, business commentaries and news reports. Invariably, each of these try to generalise the key determinants of success from the examples of certain companies. In more cases than not, those singled out as models of success falter in a few years time, sometimes victims of changing circumstance or victims of their own success.

Let us look at the current business climate in some detail. As of early 2013, the global financial crisis is a five-year-old memory. The stock market is bouncing up and down, and people are hopeful of an eventual recovery. On the other hand, the situation in Europe, particularly in Greece and other Mediterranean rim nations, is still grim.

The abysmally high level of debt remains and the capacity to repay it is elusive. On that measure, even the recovery in the United States appears suspect. China is the only bright light on the global economic horizon. And even it is faltering like the last domino.

Whatever the reasons for or the outcomes of this current global economic scenario, it cannot be denied that eventually only businesses will lead economic recovery. The failures of command economies in the erstwhile Soviet bloc have demonstrated beyond doubt that statist policies, while good short-term Band-aid solutions, can rarely lead to true economic prosperity.

Globalisation has been blamed by both the extreme right and extreme left for a host of economic ills facing various nations. Having seen its effects at close quarters in more than 100 countries, I cannot disagree more. In my view, the results attributed to globalisation are more attributable to other factors, such as human malfeasance, institutionalised corruption (even in the highest places), laziness, a sense of entitlement to riches without working for them and herd mentality leading to action without thinking and similar such factors - all part of basic human nature.

It is fair to say that one of the most alarming trends is the continued shifting of the global manufacturing capacity to China. Like a giant vacuum cleaner, China has sucked in the manufacturing capability from rest of the world at an intense pace over the past 15 years. Isolated stories of reversal of this trend apart, China continues its policy of government-aided industrial expansion far and wide beyond what might be good for its people or global economy. Its companies invest even in industries where labour provides little competitive advantage, and China has little capability advantage. This sets up a scene for an international showdown, on the economic front or on a military/political front - either of which will be an unwelcome backlash against globalisation and leading to a reversal of global economic fortunes.

In the midst of this macro situation, businesses continue to suffer from intense uncertainty and anxiety. Investment decisions are delayed for months, if not years, by the inability to project cash flows and the expected rate of returns for the investors. Waiting for consumer spending to pick up has been futile so far.

The intense volatility of commodity prices is leading to speculation and, some suspect, even manipulation in the commodity markets on a global level. At a micro level, this volatility in the prices of raw materials has made it nearly impossible to make and keep budgets, make price promises or stabilise customer-supplier relationships in a strategic manner professed by traditional supply chain management pundits.

There is no doubt that businesses large and small are suffering around the world. Hardly a business exists which was not affected by the GFC of 2008. While there has been recovery in business confidence, hiring has been sparse, investment in research and development, as well as property, plant and equipment outside of mining, has been negligible and profitability has scarcely improved. In this economic scenario everyone - ranging from business owners and executives, political leaders and the disaffected, are clutching at straws. Inevitably, all eyes are upon the business section to find a way out of the economic morass.

If business is the ladder to economic salvation, then innovation forms the rungs. New products, new methods to create, design, distribute and service these products, alongside new ways to relate to the customer and understand their world are all equally important sources of innovation. While product innovation looks sexy and garners all the attention in expositions and the media, process innovation is equally lucrative - just ask Mr Henry Ford and Mr Roy Kroc, the founder of the McDonald's franchise model.

This book should spur thinking about how to design your own business model that works for your circumstances. It is about new ways of doing things based on observations from the real world. It is not based on the study of countless subjects in a university laboratory; in real life no such laboratories exist. Neither is it based on countless interviews with executives and managers. Most executives and managers are too busy fighting the fires in their own businesses and have little time to think beyond making quarterly numbers. Those who do think in this realm, occasionally or regularly, tend to make themselves known through their writings, speeches and general interaction with academics and thinkers. Several anecdotes, as well as concepts in this book, are based on such interactions. Some of them are from client situations and business cases that I have personally worked on as a management consultant over the past 15 years and, before that, in the merchant navy.

Every day, I continue to learn and use new skills, new concepts and new ideas. In my mind, there is no higher pursuit than learning how to make things a little better one step at a time. If most of us do that, slowly bigger things such as nations, economies and even cultures, will take care of themselves.

I entreat the reader to use the concepts, materials, models and ideas presented as guidelines. As in all other situations, blind adherence to diktats or empty slogans (such as think global, act local) is neither practical nor beneficial. Take what makes sense in your situation, mould it to your circumstance and use it. For this reason, the readers will benefit from coming back to this book again and again - rather unlike most modern business books which present a single idea with a lot of anecdotes, statistics and examples.

In the first section, we will discuss the trends of networks and the value they are creating in today's society. We will see the reasons why these networks are valued so highly, despite being merely social networks of a virtual nature.

In the second section, we will examine the trends of business networks of real nature and how they are taking over the entire business modus operandi. This growing trend counters most of the traditional wisdom about how to run business. Great thinkers – from Peter Drucker to Tom Peters to Michael Porter – all dispensed valuable business wisdom. Many of the basic tenets of their thinking do not apply under the current circumstances, as you will see. Newer business models - such as Barry Nalebuff's Co-opetition-apply a lot more closely to these business networks. Collectively, they are worth several trillion dollars - far more than Facebook, LinkedIn and Twitter combined. We will see how these business networks are now configured and harnessed.

GLOSSARY

APP: Advanced Product Phasing

C2C: Cash-to-cash

FAR: Fire-Aim-Ready

IT: Information Technology

JIT: Just-in-time

PE: Price earnings

R&D: Research & Development

RFP: Request For Proposal

SCOR: Supply Chain Operations Reference

t: $eed-to-$tore

TOP: Transaction Optimisation Profitability

TQM: Total Quality Management

Section 1

NETWORKS

SECTION 1

NETWORKS

> ***All men are caught in an inescapable network of mutuality.***
>
> *Martin Luther King Jr. (1929-1968)*

Chapter 1 - The Most Important Network on Earth

- Business networks give more value than your business infrastructure.
- Business networks make your business more resilient and responsive.
- When cash is the king, business networks triumph.
- In boom times business networks give you speed.
- Business networks help smooth out the volatility - especially during times of extreme volatility.
- How does this make business networks valuable?

- If you cannot make your business networks more visible and manage them proactively, you may be silently yielding the competitive advantage to others who can.

Chapter 2 - Open Secrets from Ancient Networks

- The Silk Road - a vast ancient business network
- Venice - the powerful business network of the Middle Ages
- The Spanish Empire - an imperial business network
- The East India Company - first multi-national company
- American railroads - a business network opens up the continent
- Global finance houses - winning business networks of the 20th century

Chapter 3 - Lessons from Modern Networks

- Plethora of options in social networks are mind-boggling.
- Social networks are popular, but business networks are more lucrative.
- Basis of competition has long ago shifted from competition among companies to competition among supply networks.
- Now, the market leaders network only with those suppliers and customers who are themselves market leaders in their own chosen arena.

- Valuation of business networks ranges into trillions of dollars.
- Supply networks are rapidly undergoing massive transformations as CEOs adjust their business strategies to global realities.
- Other examples of networks abound

Chapter 4 - The Hidden Power of Business Networks

- The network effect
- Beyond synergy - the multiplier
- Synchronicity - the real power of effective business networks
- Value accretion vs. value synergy vs. value synchronicity
- Key principles that highlight the differences between the three strategies
- The three levels on a Jacob's ladder
- The journey from free markets to networks

Chapter 5 - Truth about the IT Systems Required for Great Business Networks

- Why we need a discussion of the 5-STAR Business Networks IT systems.
- Flexibility/complexity trade-off in IT systems
- Common mistakes in business network IT systems

- The mistakes encountered in the supply network information technology deployment during all four steps
- Information technology outsourcing strategy development is rarely well thought through.
- IT software and service provider selection is muddled and features driven.
- During the IT service provider implementation and integration, users frequently lose control of the project.
- In IT service provider relationship management and service usage, there is insufficient way to measure, manage and improve.
- It is important to understand the evolution and trends that have underpinned the complex business networks.
- Supply network management strategy and capabilities evolved from primitive to esoteric in the four decades since mid-seventies.
- Supply network management processes evolved in concert with the strategies and capabilities of the business.
- Software and information and technical architecture evolution underpinned the development of supply network processes, strategies and capabilities.

WHY BUSINESS NETWORKS ARE SO IMPORTANT

Chapter 1

The Most Important Network on Earth

> ***Call it a clan, call it a network, call it a tribe, call it a family. Whatever you call it, whoever you are, you need one.***

Jane Howard, "Families"

There was a time, not more than a few decades ago, if you were General Motors you would attempt to own every part of your business. The assembly lines, the parts manufacturing plants, the stamping units, the ancillary units and even the software that runs the business, the dealerships that sell the cars, the steel mills, even the mines that produce iron ore for the mills. This is for good reason - you either could not trust others to be savvy enough to produce and send you the material you wanted when you wanted it, or the margins in each of those businesses were big enough for you to try and own all those operations.

There was only one thing wrong with this picture. Your business became an insular behemoth - far removed from the customers and moving slowly in a marketplace going through a rapid transformation. Your more nimble competitors, with loose networks of aligned companies, could easily run rings around you in no time - both in terms of developing and launching new products, and producing and selling high quality products at lower prices.

Business networks are more important than your business infrastructure

Eventually, realising the truism inherent in the folk wisdom of farmers when they say you do not have to own the cow if all you want is the milk, you would investigate ways of carving out parts of your business into independent entities that could be run as a loosely aligned network of businesses, similar to what your competitors had evolved into. This is not a book extolling the virtues of keiretsu, chaebols or similarly exotic-sounding Japanese and other Asian business structures. It is, however, useful to take some lessons from the evolution and success of these business networks.

Over the past several decades, both the global economy, as well as business structures, have evolved dramatically to such an extent most businesses have no recourse but to create business networks akin to those mentioned above. So what is the magic of these business networks? Why are they so important? What makes one business network better than another? Is there a way to systematically assess, measure, report upon, improve and monitor the quality of your business network? What outcomes should you expect out of a well-tuned business network? These are some of the questions we will answer in this book.

Business networks make your business more resilient and responsive

When the tsunami flooded the eastern coastal stretch of Japan in March 2011, the ensuing nuclear disaster combined with the devastation caused by the ocean to disrupt businesses around the world. The Japanese economy sits right in the middle of the global business network and it was natural for businesses as diverse as auto manufacturing, electronics, chemicals, petroleum products, computers and metals to experience the disruptive shock. For example, the price of the Toyota Prius went up by nearly $2,400 after rumours of shortages. While it is natural for a variety of businesses to experience the disruption, it was remarkable to note that those which had the most responsive and resilient business networks were the ones to recover from this catastrophe the quickest. Later on in this book, we will see how to recognise the quality of business networks and make them more resilient and responsive at the same time.

Another stark example of the power of business networks is the fire that tipped the balance within an industry. Two stalwarts in the mobile phone industry were equally impacted by the same event - a lightning fire in the chip manufacturing plant of their common supplier, Philips, in New Mexico in March 2000. Nokia and Ericsson experienced the business disruption to an equal extent as a result. Fire damage to their stock was extensive.

More importantly, the manufacturing capacity was damaged and it was difficult to estimate the time for repairs. Nokia has invested months, if not years, in creating and perfecting a robust and responsive business network, while Ericsson's business network was a relatively middle-of-the-line affair that worked well when things were good. After the fire, Nokia was able to see the full impact of the chip shortage on its own business, as well as the entire industry, with a lot more clarity than Ericsson, or even Philips. Moving quickly, it activated other points of its business network to shore-up supplies, to redesign some of the chips to manufacture them in other plants and to take other pre-emptive steps in the network. Ericsson let the situation evolve at its own pace and made decisions more reactively. The resulting gain in profitability and market share for Nokia and the loss of these for Ericsson, tipped the balance of the industry to an extent where, within a few years, Nokia pulled far ahead of Ericsson, which never caught up with its rival.

When cash is the king, business networks triumph

This is not a unique example of a more robust business network trouncing a relatively ordinary one at a crucial time. The story is repeated in industry after industry whenever challenging times arrive, as eventually they do. In fact, the data from a systematic study carried out by the Aberdeen Group reveals two critical insights. Firstly, the companies with more robust business networks have far superior cash conversion cycles (we will discuss this in Chapter 9) - nearly six times better.

Take a look at Figure 1.1 below:

Figure 1.1

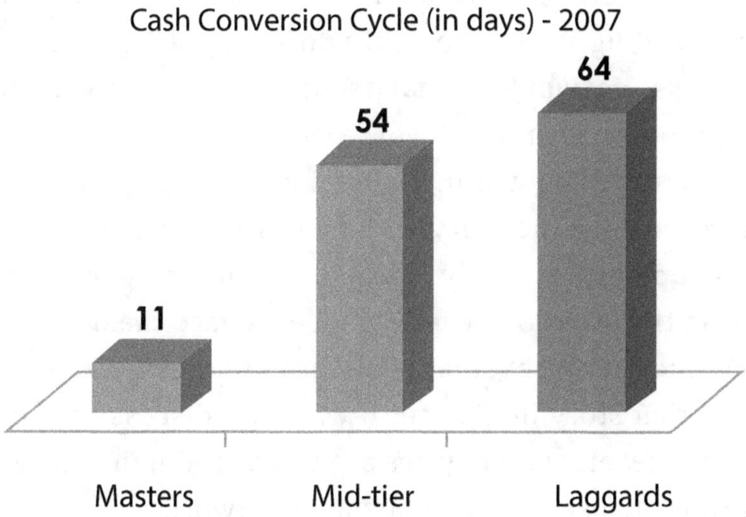

Source: http://itnewscast.com/applications/your-supply-chain-needs-be-integrated-value-chainbut-how

Aberdeen Group, Working Capital Optimisation, June 2007

Figure 1.2

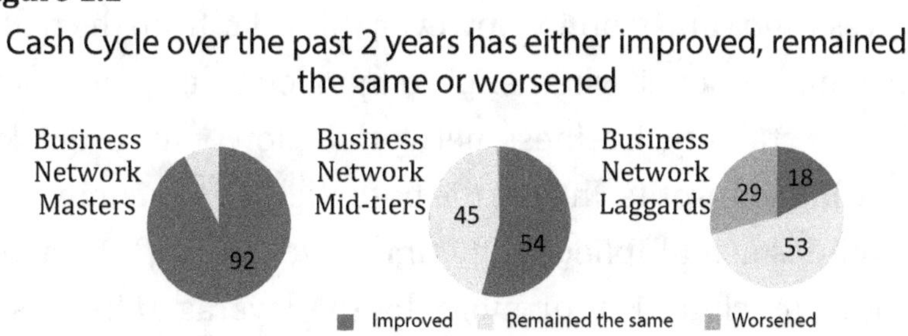

Secondly, and more importantly, their cash conversion cycle actually improved during the two years of challenging times leading up to the Global Financial Crisis (GFC), while the rest of the industry went backwards. As you can see in Figure 1.2, **92 percent of business network masters** improved their cash cycle over the two years leading up to the GFC while only **18 percent of business network laggards** had seen improvement; and 29 percent suffered worse cash cycles.

Is the result shown in Figure 1.2 surprising? Hardly, if you reconsider the story of Nokia vs Ericsson. The same catastrophic fire nearly decimated one company, while leaving the other one even stronger to face the onslaught of another heroic business network - iPhones. We will examine that story in Chapter 8 when we discuss how new product development cycles are shortened immensely by appropriate use of robust business networks.

In boom times, business networks give you speed

The above examples and data might lead you to conclude that strong business networks are useful only during times of duress and catastrophe. However, this is far from the truth. In economic booms, whether accompanied by economic volatility or stability, business networks allow you to quickly realise higher profits. In fact, the potential of your company's capabilities is multiplied many times over, perhaps by a factor of as much as 100 or more, by the leverage effect provided by your business network. We will discuss the leverage effect and other key characteristics of the business networks in Chapter 4.

Let us consider another example - the booming commodities industry. In the past 10 years, no other industry has grown as much. Prices of iron ore, copper, coal, gold, etc. have gone up exceptionally during this period. Within this industry, iron ore is one of the largest and most capital intensive operations and this industry is dominated by three global giants; BHP Billiton and Rio Tinto of Australia and Vale of Brazil. To build a large iron-ore producer and exporter from scratch, took Andrew Forrest of Fortescue Metals the full extent of his business network. To raise finances, build mining infrastructure, build and gain access to the logistics infrastructure and market the product in the international commodities trade, Fortescue Metals built a capacity in three years would have taken its much larger rivals would have taken more than 30 years to build. This was only possible through extensive use of business networks that the company built, nurtured and managed effectively. Most executives grossly under estimate the value and efficacy of business networks in ramping-up capacity rapidly in times of boom.

Business networks help smooth out the volatility

Now, let us look at another example in a highly volatile industry. Global bulk shipping is one of the most volatile industries, with the shipping rates falling by as much as 94 percent within a period of four weeks, or rising by up to 400 percent within a period of few months. In such a volatile business environment, budgeting and planning can become a nerve-racking exercise for all companies, except those which use their business networks to cushion the lean periods with long-term contracts and find scarce capacity during the boom periods.

In fact, if we examine the key success indicators through the economic cycles, a clearer picture emerges, as shown in Figure 1.3.

Figure 1.3: Basis of Competitive Advantage

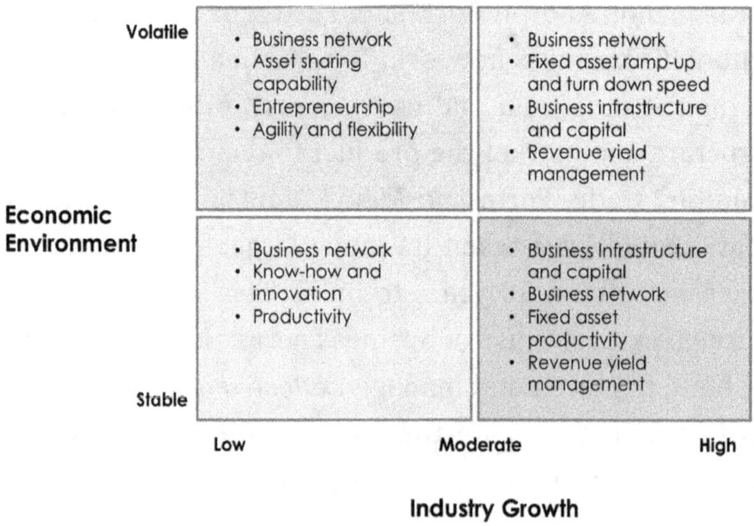

We will see these examples, and many others, in a lot more detail throughout the pages of this book. Suffice to say here that the business networks quoted in the above examples, also called the supply chain in many instances, are unique to every company and industry, formulated over a number of decades in many cases, and are worth several trillions of dollars in value. In the retail industry alone, the entire business network is worth more than $1.5 trillion according to one estimate. Now, contrast that with the much-heralded valuation of Facebook, currently pegged at more than $80 billion, to put things into perspective. Obviously, social networks are not the same as the business networks. We will compare and contrast these two in Chapter 3 to understand what they can learn from each other.

Valuation of business networks is astronomical, though difficult to quantify

The total valuation of the business networks in just six industries - retail, automotive, electronics, FMCG, aerospace and mining - is estimated to be more than $4 trillion. These hidden business networks are reformulated and reconfigured on a continual basis - sometimes leading to massive transformations on global scale, while at other times leading to gradual transformations that only manifest fully over a decade or more.

Mark Zuckerberg is obviously the hero of the social networking world. Tonnes of ink has been spilt writing about his exploits in the social networking scene. So, who are some of the heroes in the business networking scene? How about the likes of Michael Dell, Steve Jobs, Dietrich Mateschitz (founder of Red Bull) for some obvious examples. Michael Dell reconfigured the entire business network in the electronics and computers industry to reduce the cash conversion cycle of his company from 56 days which was the industry norm, to -4 days over a period of four years. We will talk about the Dell story in a lot more detail in Chapter 9.

Let us now look at Steve Jobs' role in business networking. His, and Apple's, exploits in new product development are well-known and highly regarded. However, what is less well-known is the role of the business network of suppliers, collaborators, partners and other such entities who have aligned their own interests with that of Apple, in order to reduce the new product development cycle to nearly eight months, from an industry norm of approximately 26 months. Again, the magic of business networks has made it possible to design, build, launch and sell revolutionary products in less than one-third time of the industry. We will examine the role of business networks in new product development in detail in Chapter 8.

Finally, to wrap up this chapter, let us examine the role of an unlikely business networking hero - Dietrich Mateschitz, the founder of Red Bull. Mateschitz, a marketing genius, rolled out his drink Red Bull in more than 100 countries in less than 17 years – a feat that took its much larger rival nearly 69 years. How did he manage to do that with a much smaller capital base and a smaller product range? The answer lies in his vision and business networking skills to formulate supply chain outsourcing relationships rapidly and manage these effectively. We will examine the role of business networking in modularised outsourcing design in Chapter 12.

These are just some of the most well-known examples in the realm of successful business networks. A vital question arises at this point. What are the differences between the successful and not-so successful business networks? What distinguishes the business networking masters from the mid-tier players and the laggards? After all, the whole point of reading a book is to get some practical and actionable advice on how you can ascertain the nature your own business network and take steps to improve it, if you deem fit. We refer the super networked businesses as having a 5-STAR network and will compare and contrast the masters and the rest in Chapter 6.

So fasten your seat belts, come with us on a journey through the business networks around the world and see how successful businesses are configuring the business networks of tomorrow to thrive through the economic gyrations of the 21st century. It will be a worthwhile journey, not only for broadening the horizons, but for profiting from experiences of other masters of the game.

SUMMARY:

Business networks are more important than your business infrastructure. Typically your business infrastructure is static, rigid and cost accruing. Your business network, on the other hand, is evolutionary, flexible and revenue accruing. Business networks make your business more resilient and responsive at the same time. When cash is the king, business networks triumph. Data is clearly conclusive that, in times of cash crunch, the quality of their business networks help save companies.

Besides, in boom times business networks give you speed - even relatively smaller businesses can achieve remarkable results quickly based on the responsiveness of their business networks. Business networks help smooth out volatility - especially during times when it is extreme. No wonder we see airlines and shipping companies forming global service alliances to ride out the seasonal and economic peaks and troughs. Valuation of business networks is astronomical, though difficult to quantify. Estimates range into trillions of dollars, may still be underestimating the full extent and power of these hidden business resources. If you cannot make your business networks more visible and manage them more proactively, you may be silently yielding the competitive advantage to others who can.

Business Networks
Through the Ages

Chapter 2

Open Secrets from Ancient Networks

> ***Plus ça change, plus c'est la même chose.***
>
> *Alphonse Karr, Les Guêpes 1849*

Before we embark on the journey of exploration into the current realms of business networks, let us take a journey into the past. A logical question always is whether this is new or has it been done before. After all, as the quotation above says, the more things change the more they remain the same. This is true of the business networks too.

In a way, the networks we are talking about in this book are not new at all. Some features, such as speed, trust-building mechanisms, outcomes and science of network creation and harnessing are new. However, similar networks have existed in formally or informally through the ages. And each time, they have been very effective in creating value for the participants, business community, and society in general. In this chapter, let us briefly visit some of these networks through the centuries and see what we can learn from them. History is our best teacher.

At this point it is appropriate to note that we will keep these discussions of historical business networks brief, neutral and non-controversial. The reason, besides efficiency, is that our objective is not a detailed study of characteristics, pros and cons, or even moral rectitude of the participants. Rather, it is the more mundane exercise of finding parallels with the current business networks and their key characteristics.

The Silk Road - a vast ancient business network

Without going too far back into antiquity, let us start our journey at the Silk Road of the Middle Ages. At its peak, the Silk Road extended nearly 4,000 miles from the coast of China through the territories of China, Central Asia, India, Persia and Assyria to the Mediterranean coast of Levant and onwards by sea to the fabled city of Venice and beyond. The term Silk Road creates an image of a road, perhaps even a boulevard traversed by a camel train laden with expensive cargoes, well protected by security. The truth is, however, far more complex. In reality, it was a complex network of caravans, camel trains, serais, traders, money lenders and ships extending across the known world of the time to carry merchandise as diverse as silk, spices, wool, fabrics, tea, porcelain, carpets, ivory and other items of high value. Each of the main cities in this network - Kashgar, Samarkand, Turfan, Baghdad, Tyre, Aleppo and Alexandria - was a veritable hub of activities related to trade facilitation. Participants in this network - whether a Sugd merchant, a Chinese caravan owner, or a Florentine ship captain - were handsomely rewarded for their enterprise and their ability to work their part in harnessing the power of the network.

Let us look at how the network was organised in practice. In the agrarian society of the Middle Ages, the most precious commodity traded by the business network was spice. Before the advent of refrigeration, spices were necessary to preserve food, as well as mask the flavor of spoilt or rancid ingredients. Many spices, such as pepper, cloves, mace and cumin, ginger, nutmeg, cinnamon and saffron, traded for margins up to 3000 percent and some were regularly sold for prices more than their weight in gold. With such high margins, demand was never a constraint, the extent and security of the supply route was. It was usual for the goods to change ownership more than 25 times between the producers in the East and the eventual consumer in the West. Each middleman added his own margin, as well as embellishment to the story. For example, Arab merchants told their European buyers that the cloves were netted out of the river Nile and cinnamon came from the birds.

Without any central command, the entire network was organised towards one end - production, collection and transportation of the spices and similar produce to destinations in Europe. Numerous middlemen, camel caravan owners, merchants, ship owners, and financiers, participated in this chain - each within their own territory and with their own margin. It is estimated that the middlemen's share of the profit in this enterprise was more than 90 percent. [1]

Venice - the powerful business network of the Middle Ages

No wonder, then, that the Venetian merchants - the middlemen closest to the end customers and with the most visibility of the entire network - moved to take greater control of the network. In the late Middle Ages, the bankers, merchants, and ship owners of Venice controlled the trade into Europe. At the peak of its power, the Venetian Republic had a fleet of ships exceeding 3000 vessels and controlled all trade from the eastern shore of the Mediterranean Sea to Europe. To assure security of the supply network, the Venetian Republic controlled territories on the Adriatic coast, so pirates could not attack their ships coming from the east. Using their financial power, the Venetian Merchants' Guild began to dis-intermediate the middlemen in the entire supply chain. The simultaneous rise of strong Arab empires with central command, first in Baghdad and later in Cairo controlling a vast territory, aided this process of dis-intermediation as the caravans could traverse a much greater distance and were assured relatively more security in their travels. A key feature of the Venetian business network was the role of political power, as well as finance, in shaping desirable business outcomes and securing these outcomes for the key participants in these networks. Evidently the network had become a lot more formal in its form and functioning (or possibly more records are available to give us an impression that it had become more formal). It should not surprise us that consolidation of political power made the business network more effective and efficient at the same time. There are ample historic records, as well as anecdotal writings - from Shakespeare's "*The Merchant of Venice*" to Horatio Brown's historical reviews - that make great reading about the business network centered around the Venetian merchants of the later Middle Ages. [2]

The Spanish Empire - an imperial business network

In the pre-Renaissance period, spice was still the commodity in demand and the margins were still extraordinarily high. Despite dis-intermediation of many middlemen, the land route, with its camel trains and shifting political fiefdoms, was always the second-best alternative to a sea route that directly accessed the sources of supply. With this in mind, Christopher Columbus set out in search of the sea route to India and discovered America instead. However, another explorer, Vasco da Gama did manage to open the sea route to India for Portuguese traders. The business network over the next two centuries came to be dominated by Spain and Portugal - trading with the new world of the Americas, as well as the old world of Asia. The key feature of the business network of this era was a complete merger of political and economic action in the name of the Sovereign and the rise of colonialism as a result. An exploitative business network of this nature, with its one-sided value exchange, could not last very long. At one time Vasco da Gama captured the local king's emissary, who had earlier afforded him a grand welcome during his much celebrated first visit to Calicut in May 1498 - called him a spy, ordered his lips and ears be cut off and after sewing a pair of dog's ears to his head, sent him away. This network was gradually replaced by a more benevolent form of colonialism - the East India Company and its peers. [3]

The East India Company - the first multi-national company

The East India Company started trading in spices and gradually took control of almost the entire spice network. As it expanded its reach through political and business alliances and skirmishes, it established beacheads in trading centres around the world and extended the trade itself. Tea was cultivated in Ceylon and India for export to Europe. Opium was cultivated in India for export to China and silk and silver were exported from China to Europe. With the advent of the industrial revolution, the business network become more diverse - covering greater areas and a larger number of commodities - as well as more efficient, as faster steamships and railways replaced the sailboats and camel trains. Other, less dignified trades, including those in slaves and opium, became more lucrative and perhaps reduced the brain power invested in more legitimate trades, such as industrial goods and raw commodities. While the pros and cons of the rule of the East India Company are well debated, we will restrict our discussion to the business impact of the network created by the company. The key noteworthy feature of this business network was the use of alliances to establish the network, and use of technology to improve efficiencies in its results. [4]

American railroads - a business network opens up the continent

As the business network of the East India Company was getting embroiled in exploitative trades of opium and slaves, another network, far more effective, was being established on the continent of North America. Having won the American War of Independence, the colonies of the eastern seaboard were expanding far into the continent of North America - in all directions. When the technology of railroads eventually arrived, this expansion boomed. With the rapid industrialisation and growth of railroads, it became possible to mine, establish factories and build all the modern transportation and communication networks that underlie modern commerce today. Besides opening up the entire continent and establishing the United States of America as the pre-eminent industrial power of the 20th century, the railroad network also formed the basis of massive fortunes of tycoons such as Stanford, Pullman and Vanderbilt.

Oil, steel, coal, chemicals, power stations, mining operations, grain and passenger transportation were all greatly boosted by the railroad network. It is fair to say these formed the basis of the modern industrial economy of the USA. The key feature of this business network was the clear enunciation of the positive reinforcement cycle, or flywheel effect, whereby other industries got a boost from establishment of the railroad network, which in turn further benefited the railroad networks themselves. [5]

Global finance houses - winning business networks of the 20th century

The last business network we will consider in this chapter is the network of finance houses of New York, London and Frankfurt that gained strength in the 20th century. With massive industrialisation came the need for massive financing of the industry. At the same time, growing world trade needed the finance network to be global in nature. Imagine an oil trader in Singapore buying oil from a Russian producer and selling it to a buyer in China for use in Chinese automobiles. The trader has no way to ensure that the seller will deliver him exactly what he purports to offer and that the buyer will pay for the oil on time - unless the modern finance house steps in with supply chain financing solutions, such as a humble letter of credit. From such simple finance instruments, the modern world of commerce and industry finance has developed into complex and bespoke financial arrangements that suit the needs of investors, commerce, industry and speculators. Facilitating all these transactions, and keeping trust in the system, is a network of businesses involved in the business of finance. Without going into the history of the finance houses, such as Barings, Rothschild, JP Morgan and others, we will take the key lessons from the outstanding success of these business networks. Undoubtedly, the first reason for their success is the value they create for their customers, since no business can thrive for any length of time without that. This is underpinned by their ability to understand a diverse set of business circumstances that their own customers face and to readily come up with flexible solutions to meet those circumstances. Ability to attract, train and retain talent is the key to understanding your customers. Patience, ability and willingness to shape legislature, ability to formulate and enforce a trust-building mechanism in the network and a willingness to work through the ups and downs of the business cycles, are some of the other drivers for success. A key feature enabling the success of this business network is the overwhelming practical nature of the participants, without any dogmatic adherence to outdated ideas. To the extent this network retains its ethos of service and humility, it will continue to prosper and retain its relevance in the coming era of simultaneous globalisation and fragmentation.

SUMMARY:

Business networks have thrived through the ages in various forms. In this chapter, we briefly examine six relatively well-known business networks from the history. The Silk Road - a vast, ancient business network - comprised hundreds of sub-networks, each functioning on its own, in order to achieve a vastly superior business outcome. Venice - the powerful business network of the middle ages - started the trend towards consolidation of power for achieving business outcomes. The Spanish empire - an imperial business network

- took this trend to extremes, where it became impossible to distinguish between the business goals and the imperial goals of the network. The East India Company - the first multi-national company - established commercial networks that still thrive after more than 250 years. American railroads, a business network that opened up the continent to thriving commerce and led the industrial revolution in the new world. Global finance houses, winning business networks of the 20th century, funded the industrial Revolution and benefited from its vast reach.

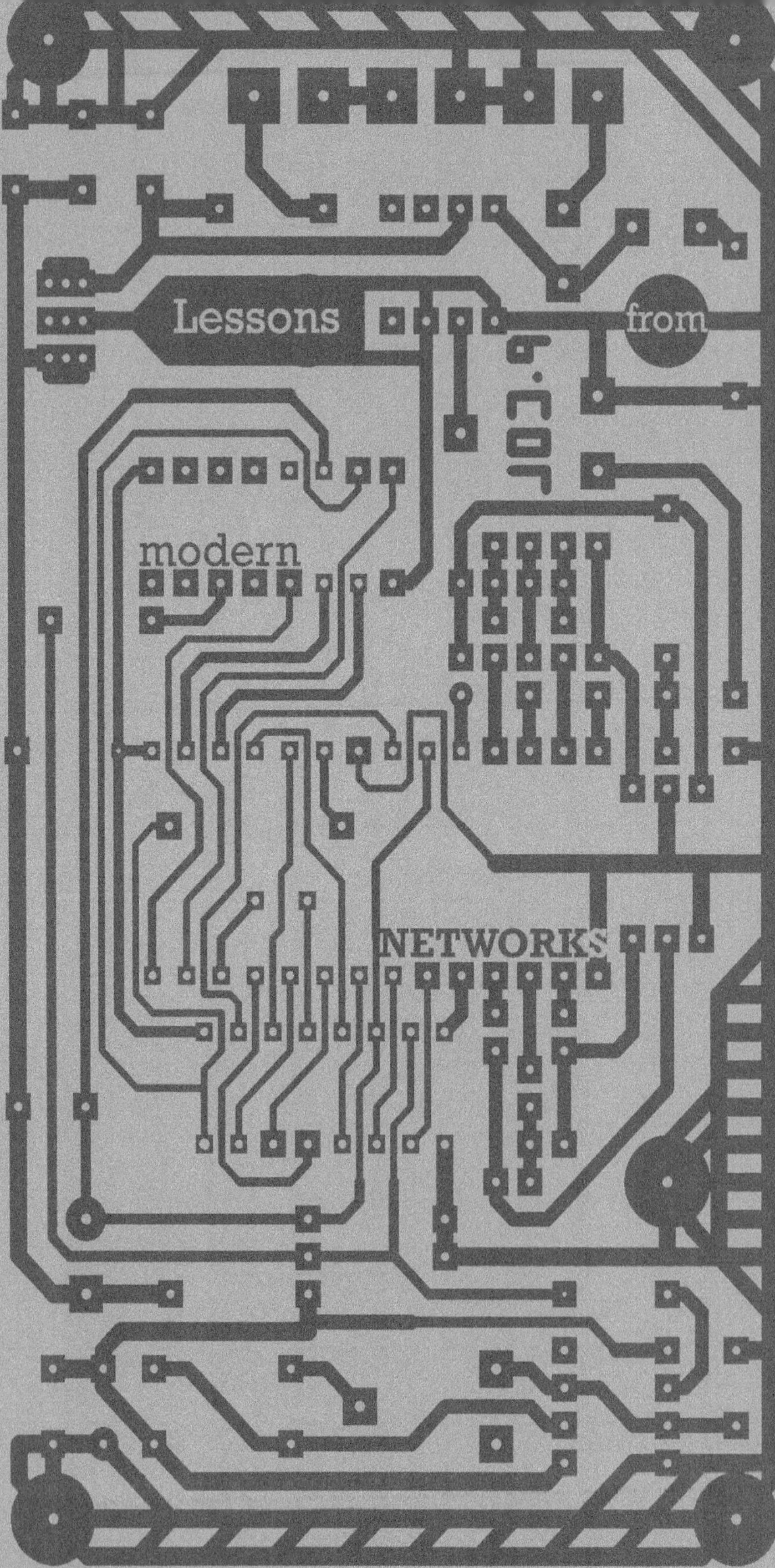
Lessons
from
modern
NETWORKS

Chapter 3

Lessons from Modern Networks

> ***Every man takes the limits of his own field of vision for the limits of the world.***
>
> *Arthur Schopenhauer (1788-1860)*

Let us cast our field of vision on modern networks, which come in a vast array of structures and forms. Depending on the value they provide to participants some of these networks are successful, while others are quickly replaced by more efficient models.

In this chapter, we will explore key distinctions between the various types of networks and their respective strengths and benefits. In Chapter 5 we will take an intensive look inside the IT systems used by business networks. This chapter is an extensive exploration of the networks themselves. Many of these networks use the kind of IT systems we will examine in Chapter 5.

Plethora of options in social networks are mind boggling

When researching the social networking scene, I was aghast to find the plethora of options depicted by the following chart shown in Figure 3.1 on the next page.

Obviously, Facebook and Twitter are the dominant gorillas on the social networking scene. Whole gaming networks and other apps have cropped up to service the participants on Facebook and Twitter. A plethora of other structures, such as social brand engagement, social marketing networks, social intelligence, social scoring and social referrals are cropping up. Most of them are too new to discern which models will eventually survive. However, in this clamour, one thing is abundantly clear - social networking itself is revolutionising the way humans live, work, interact and govern. While the Arab Spring in Tunisia, Egypt and many other countries is now a fading memory, the role of social networks in creating impetus and sustaining momentum cannot be underestimated.

Social networks are popular but business networks are lucrative

In this book, our focus is on business networks. We will not spend much more time on the social networking phenomenon; powerful, diverse and exploding though it is. A far more powerful, diverse and well-established network exists in the business arena. Every business relies on a network of suppliers and customers to sustain its operation and place in mod-

Figure 3.1: Universe of Social Networks

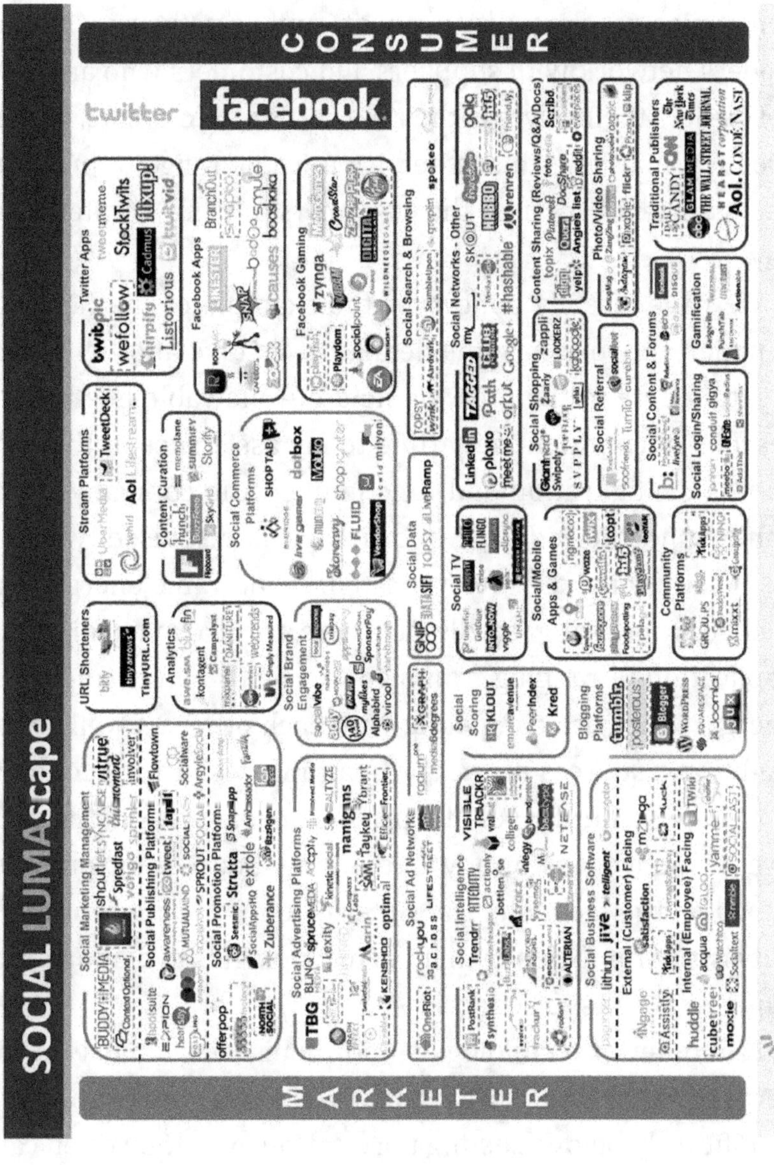

ern society. So powerful is this network, in fact, that the basis of competition has long ago shifted from competition among companies to competition about supply networks. Market leaders network with suppliers and customers who are market leaders themselves in their own arena to co-create, design, produce, market, sell and service a whole range of products and associated services the customers take for granted each day.

We have already gone over the example of the Red Bull supplier network in some detail. Let us take another example, this time in the growing field of renewable energy. For reasons of confidentiality (due to my assistance in this network's configuration and functioning), the company is not named and the industry information is disguised, without losing any context or relevance. The renewable energy field is not only growing rapidly, but there are also many different applications - ranging from geo-thermal to wind and solar energy, which are competing with each other, as well as the traditional sources of energy. In a rapidly shifting economic landscape, with the volatile pricing of coal and oil, the relative attractiveness of any one energy source not only depends on the technological superiority of the concept itself, but also on rapid deployment of technologies to demonstrate superior results in action.

A number of technologies find themselves almost on the verge of major breakthrough that will lower the cost per unit significantly below existing norms. However, they often cannot achieve those breakthroughs on their own because of the research and development time and investment required.

On the other hand, there exist other well-established businesses quite capable of providing additional research and development capability, even though renewable energy is not their area of expertise and interest.

In such a scenario, the challenge emerges: How to create a network of businesses with complementary capabilities that can bring in the required expertise in research and development and use it to co-create a product or technology, where each business is also satisfied while optimising its resource usage in pursuit of its core mission.

Almost all the companies in a similar scenario create a complex business network of suppliers, collaborators and co-creators that can rapidly form and deploy teams for the purposes of research, development, testing, prototyping and deployment of technologies. If any company was to rely on its own resources using traditional purchasing practices, the time to market and effectiveness would be vastly diminished. A concrete example we will examine in greater detail in a later chapter, Apple reduced its new product development and time-to-market cycle to nearly one-half of the industry norm by using its supply networks wisely. No wonder Apple's current market capitalisation is larger than that of Greece and Spain combined.

Valuation of business networks ranges into trillions of dollars

How big are the business networks we are discussing? One estimate puts the combined worth of these supply networks into trillions of dollars. The following figure (on page 39) shows the estimates for only a handful of industries:

Each industry has a highly **unique** and **valuable** supply network that its participants have created in response to the circumstances, regulations, customer needs, and economic situation in the industry.

Either by trial and error or by intelligent design somewhere along the way, each of these networks has **evolved** over the past several decades to reach into its current state as a result of all the changes in economics, technology, regulatory requirements and customer tastes. Sometimes a new player, such as Red Bull, enters the industry and causes a major disruption to the established supply relationships and networks. In most cases, the established supply relationships and networks in every industry will go through a gradual evolution as the environment evolves. Most CEOs are now familiar with the power of supply networks transformation. Currently in many industries these supply networks are rapidly undergoing **massive transformations** as CEOs adjust their business strategies to global realities.

We will cover business networks and supply networks in great detail in Sections II and III. For this reason, we will devote the rest of this chapter to looking at some other types of networks familiar to you.

The most familiar network to you is your own. A network of business associates and contacts. Inevitably, they come to your rescue when you need help and vice versa. They are also good sources of information, job leads, business leads and even guidance or inspiration. LinkedIn and Plaxo have put the Rolodex of old online and added far more functionality and possibilities

Figure 3.2: Valuation of business networks

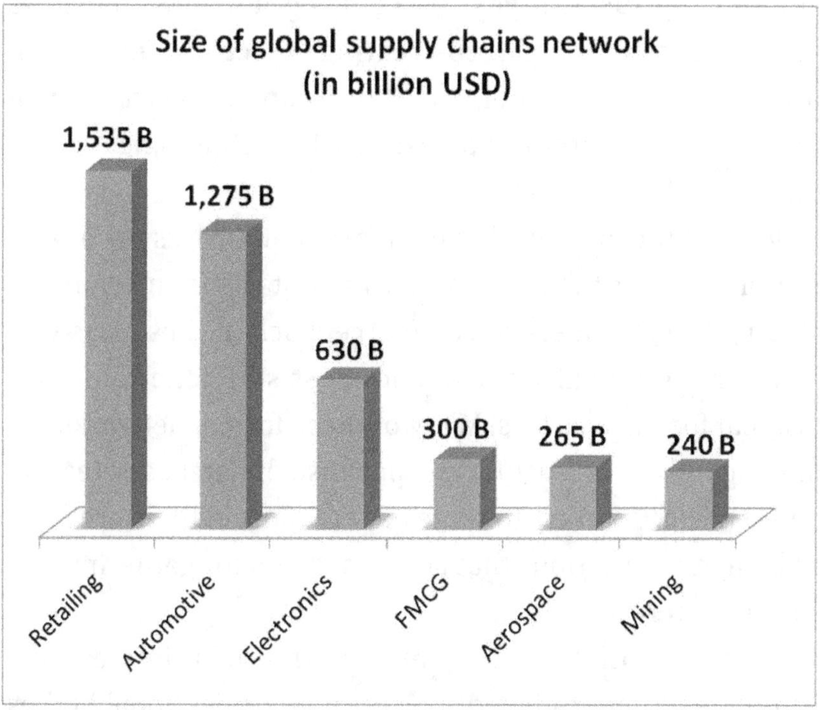

Source: Analysis by Global Supply Chain group, annual reports of 80 companies

with their web versions. In that sense, online networks enhance the possibilities and functionalities far beyond what is possible with the offline networks. An enriching, quasi-structured and purpose oriented interaction is possible using online networking tools - it would be difficult to replicate this (and adhere to it) in an offline context.

Other examples of networks abound

Other examples of offline networks are alumni networks from your schools, colleges, previous work places and similar locations. Many of these are now online, though very few are doing much beyond basic database creation and sharing. Just this year, one of my graduate schools has started systematically co-creating executive education programs with the help of its alumni network - even though the possibility has existed in technological capability for almost a decade.

Some of the more well known formal networks are organisations such as the Freemasons, the Rotarians and clubs including the chambers of commerce. Each of these was created for a different mission and most still adhere to their original formal charters. Many of these formal networks are moving online, at least for the purpose of database creation and sharing. Contrary to some beliefs, most of them are rather benign associations that create a forum for gathering and sharing interests.

Many networks of professionals with similar interests exist online either at LinkedIn or through professional bodies where the only condition of entry is your interest in the topic of discussion. One such offline network where I served on the Global Advisory Board broadened its membership by tens of thousands during the five years I was on the board. The professional courses, seminars, forums and lecture series run by the organisation contributed significantly to the body of knowledge on the subject. At the same time, numerous individuals found suitable jobs and professional progression advancement though the network.

Another online informal network I have created on a social networking website has grown to more than 6,500 members globally and business opportunities worth more than

$650 million are offered every year. It is difficult to estimate the actual trade that results from the network as trade is mostly carried out offline. This network is growing rapidly. Now I am grappling with the question of how to keep its informal nature and yet enhance its usefulness and relevance to the participants of the network.

An interesting phenomenon that is just emerging is online currencies being traded between the informal networks. These go far beyond online payment and clearance systems such as PayPal and serve as crypto-currency. So far these crypto-currency networks including Bitcoin and its clones are operating near the legal fringes of most economies. Further developments are being watched with interest by the central banks, supra-national bodies and the political economists. We will draw the line here for our discussion of semi-underground networks because one can go too far in researching and describing these and stray into illegal or unethical networks quite unknowingly.

Not all networks are benign

However, in addition to the online and offline networks more informal networks exist. Some are even sinister, the most notorious of them being Al-Qaeda. Some of these informal networks are even online where the identities of the individuals are not fully known, like the hacker group Anonymous.

CONCLUSION

In conclusion, I would like to point out that the breadth, as well as the depth of a network varies according to the need and purpose of the network itself. Most networks - whether online or offline, formal or informal, social or business - are evolving slowly to meet the needs of the participants, and continue to make themselves more useful by adding depth of interaction and/or breadth of coverage.

In the rest of this book, we will focus primarily on the business networks, their efficacy - efficiency and effectiveness and their utility to the participants. More specifically, we will demonstrate how the business leaders of the future are busily building networks that will underpin their business aspirations. There are five key leverage points of the business networks and we will discuss these in details in the next section.

SUMMARY:

Plethora of options in social networks is mind-boggling. Social networks are popular, but business networks are lucrative. Basis of competition has long ago shifted from the competition among companies to competition about supply networks. Market leaders network only with those suppliers and customers who are market leaders in their arena to co-create, design, produce, market, sell and service a whole range of products and associated services that the customers take for granted on a daily basis. Valuation of business networks ranges into trillions of dollars. Most CEOs are now familiar with the power of supply networks transformations in many industries. These supply networks are rapidly undergoing massive transformations as CEOs adjust business strategies to global realities. Other examples of networks abound and range from beneficial to sinister networks that operate offline or online. Our understanding of such networks is still evolving.

What makes the business networks great?

Chapter 4

The Hidden Power of Business Networks

"Confidence is the affective basis of action...; trust, of co-operation; and loyalty, of organisation."

Professor Jack Barbalet (1996)

In 2012, when Facebook's IPO was being discussed in the media, a range of valuations was put forward by experts between approximately $50 and more than $100 billion. Most people in traditional businesses were stunned and asked how could a company with no products, no factories, no customers and no suppliers, be valued more than Siemens, Nokia, US Steel, or even a combination of these traditional, well-respected companies. The pundits declared the basis of valuation as the Network Effect and left it at that - leaving people to decipher what exactly the Network Effect is and exactly how does it lead to a valuation of tens of billions of dollars.

In this chapter, we will not justify or challenge the valuation of Facebook, mainly because we are not in the business of predicting the future. We are merely seeking to understand the present, in order to move towards the future in a more confident manner. Confidence, after all, is the affective basis of all action.

Networks, of course, can be homogeneous groups of similar people, such as those who are net-savvy, with spare time and willingness to share their lives' details with others on Facebook (or similar social websites), or they can be heterogeneous networks of a multitude of suppliers around the world that provide a vast range of components and parts, such as those that go into manufacturing the Airbus A380. Networks can even be a combination of homogeneous entities and heterogeneous groups, or vice versa. This point should emphasise in the reader's mind that there is not one single kind of network, and hence the characteristics will vary accordingly.

The Network Effect

Needless to say, there is no point spending time on a social network website if you are the only person who ever visits it. In fact, if most of your friends are members of a rival social networking site, you would eventually find yourself there, or find yourself a web outcast. This leads to a catch-22. Most nightclub and restaurant owners are long familiar with the predicament - the more popular your establishment becomes, the more people want to get into it. However, the key predicament is always - how to start off the process. An excellent book "*The Tipping Point*", by Malcolm Gladwell,[1] discusses this phenomenon in great detail and attributes it to three rules - getting the first movers (law of the few - the mavens, the connectors and the salesmen), the stickiness factor (simple ways to make things memorable) and the power of context (small factors in the environment and the relationships that create and sustain impetus). If you have not yet read this book, I highly recommend it. It is neither advisable, nor possible, to paraphrase the excellent content, and Gladwell's writing style is extraordinarily eloquent.

So, with every new addition to a network, it becomes a little bit more valuable. This continues to happen until the network reaches a tipping point, a point at which it suddenly becomes a lot more valuable. After this point, the network will generally race past all its rivals and become a de facto standard in its realm. Whether it is a question of which social network website to frequent, or which type of keyboard to use as a standard (the more popular QWERTY or the more efficient Dvorak style), or which type of cooling systems to use in the nuclear power plants (light water, heavy water or gas cooled) - the decision almost always rests on the network effect.

To see the network effect in practice, let's use an example from the distant past, when automobiles were being first introduced. In their excellent book "*Thinking Strategically*" authors Avniash Dixit and Barry Nalebuff [2] quote the following example by Stanford economist Brian Arthur:

> *In 1890 there were three ways to power automobiles - steam, gasoline, and electricity - and of these one was patently inferior to the other two: gasoline. ...[A turning point for gasoline was] an 1895 horseless carriage competition sponsored*

> *by the Chicago Times-Herald. This was won by a gasoline powered Duryea - one of the only two cars to finish out of six starters - and has been cited as the possible inspiration for R. E. Olds to patent in 1896 a gasoline power source, which he subsequently mass-produced in the "Curved-Dash Olds". Gasoline thus overcame its slow start. Steam continued viable as an automotive power source until 1914, when there was an outbreak of hoof-and-mouth disease in North America. This led to the withdrawal of horse troughs - which is where steam cars could fill with water. It took the Stanley brothers about three years to develop a condenser and boiler system that did not need to be filled every thirty or forty miles. But by then it was too late. The steam engine never recovered.*

The authors go on to state that while there is little doubt that today's gasoline technology is better than steam, that's not the right comparison. How good would steam have been if it had had the benefit of 75 years of research and development? While we may never know, some engineers believe that steam was the better bet. It is indeed a fascinating story of how the Network Effect helped a relatively inferior technology triumph. Interested readers can read a more detailed account in the publications quoted above. My purpose is to give enough details to keep the narrative tight and fast, and not send you on wild goose chase. I encourage you to follow up on this link if you need details.

So, how can networks achieve what individuals cannot? Why are networks even more valuable than similar-sized companies and organisations? What distinguishes networks from teams? What is the secret source of networks? These are some of the key questions that must be answered in this chapter.

Beyond Synergy - the multiplier

At the start of this chapter we quoted Barbalet saying confidence is the affective basis of action - particularly individual action, since confidence is an individual characteristic. Similarly, mutual trust is the affective basis of all co-operation - particularly in a team. Co-operation creates that elusive thing called synergy that makes it possible for 2 + 2 to become 5, and is so highly sought after in mergers, acquisitions and takeovers. It is the basis of payments over the replacement value of the property, plant and equipment. But networks have something else working for them - far over and above the synergy - that almost makes 2 + 2 into 22. What is it?

The classic example to illustrate how the Network Effect builds up is the story of fax machines. It is an often articulated truism that the first fax machine in the world was very lonely. In fact, the first few sales of fax machines would have been next to impossible, not just because of the immense costs of the machines themselves, but because after all, there was no one who could receive the faxes. The second machine increased the utility for both - the first and the second machine - by a factor of two, because now two potential strands of communication were possible. The third machine, however, did not increase it by three - it increased by a factor of 3 x 2 = 6. The fourth machine increased it further by 4 x 3 = 12. In general, the utility increased by the formula N^2-N, where N is the number of participants in the network.

Figure 4.1: As the network expands, its utility multiplies

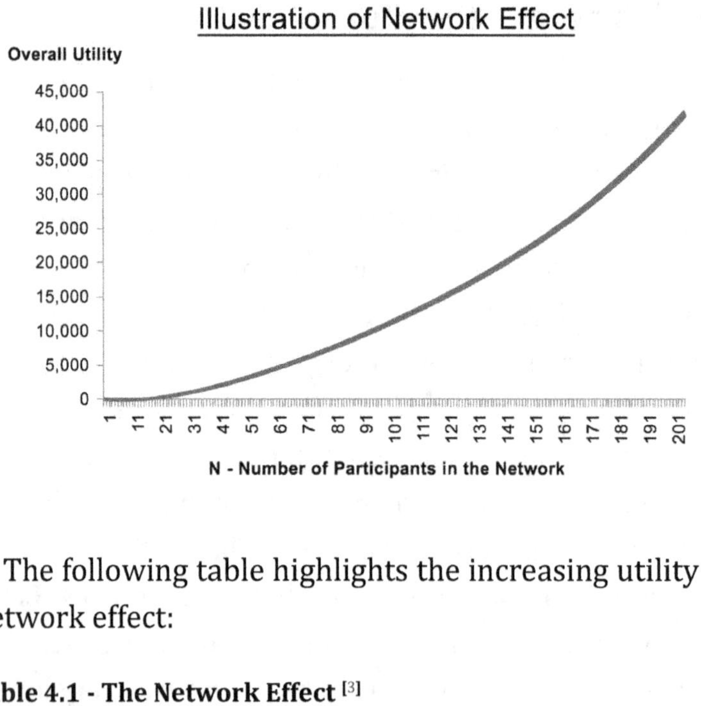

The following table highlights the increasing utility of the network effect:

Table 4.1 - The Network Effect [3]

N	1	2	3	4	5	6	7	8	9	10	11	12	13	14	15
Utility	0	2	6	12	20	30	42	56	72	90	110	132	156	182	210

As seen in the table, the overall utility increases quite rapidly and shows no sign of tapering off. In fact, Figure 4.1

highlights a similar trend when plotted up to N=200.

The question still remains - what is it beyond the confidence, the trust, the synergy, that makes the networks so valuable and powerful? Good networks already incorporate all of the above - confidence, trust and the synergy. But they have something beyond these. To explore that source, let us see an example.

Synchronicity - the real power of business networks that work

I could, of course, use the example of so many of my past friends, class mates and batch mates who I have been able to trace through Facebook in the past few years and have been a source of joyful nostalgic memories, but I will use a far more utilitarian example from my professional network built using LinkedIn.

A few years ago, when starting to write my first book on Green Supply Chains, I faced a seemingly insurmountable problem. While I already had most of the material and content from my previous work, I did not know how to shape it into a book for publication. More importantly, I had no knowledge about how to market the book to the publishers, negotiate with them, persuade influential people to read it and give suggestions for its improvement and finally provide endorsements for the dust jacket and marketing of the book.

However, I was aware of a new online professional network that was just starting to become popular - called LinkedIn. I joined a few relevant professional groups and started making contacts, asking questions and discussing the ideas. Very soon, I had found an excellent co-author (Stuart Emmett), more than 40 people who agreed to read and provide feedback and suggestions to improve, and more than 20 people who agreed to provide suitable case-studies for inclusion in the book. What is ironic is that nearly everybody who agreed to contribute kept their commitment - in full and without much prompting. In my professional career of running large-scale strategic transformation projects, I have not seen this amount of buy-in and commitment from the highly regarded and even more highly paid professionals who usually work with me on our projects. I can truly say that the professional network provided and nurtured by LinkedIn was primarily instrumental in finishing and publishing the book and its relative success. We will see many such examples of success deriving out of networks in business and economy. The key ingredient contributed by the networks goes much beyond the simplistic concept of synergy; I call it Synchronicity.

Readers who are familiar with Carl Jung's work will be gratified to see his contribution finally acknowledged in a book about creating practical business outcomes. First articulated by Jung in the psycho-analytical realms, the concept of Synchronicity is frequently used in metaphysical and pseudo-physical context. In our discussion here, we will take only some of the meanings which Jung attributed to the term Synchronicity and compare and contrast them with both synergy and its poorer cousin, accretion. Our intent here is to create a platform for showcasing the magic inherent in good networks and discern the reasons why they are so valuable.

Value accretion vs value synergy vs value synchronicity

As I think about this topic, research it and jot down my notes, I am overwhelmed by the power of the idea itself. It appears that a whole new book would be required to do full justice to this topic alone. However, here we will stick to our original intent for this particular book, which is very practical and business oriented; I will use a tabular format to summarise my thoughts, and follow the table with a brief discussion of the key points. Readers who want to discuss and develop these ideas further are welcome to contact me using the website or email address provided.

While researching this topic, I noticed a few academics and thinkers making a vague connection between synergy and synchronicity - which is quite heartening. However, the distinctions between the two concepts are neither clear nor well-articulated. Clearly, such a distinction is required and the following table attempts to that:

Table 4.2: Value Addition (Strategies)

Value Addition	Accretion	Synergy	Synchronicity
Mathematical Principle	2+2=4	2 & 2 = 5	2 & 2 is 22
Physical Metaphor	Mechanical (Newton)	Quantum (Max Planck)	Relativity (Einstein)
Cognitive Principle	Impulse	Conscious	Sub-conscious
Organizational Metaphor	Chronometer	Orchestra (hear the well - arranged and rehearsed music)	Schools of fish (see the spontaneous music in action, changing every second)
Visual Metaphor	Torch	Laser	Sunlight
Key Co-operation Mechanism	Control (Forced)	Co-ordinate (Labored)	Co-create (Spontaneous)
Economic Metaphor	Free Markets - Invisible hand	Managed Markets - Firm hand	Networks - Joined hands
Economic Principle	Scarcity	Prosperity	Abundance
Key Actors	Individuals	Organizations	Networks
Economic Guru	Adam Smith	J. M. Keynes	Barry Nalebuff
Stance	Thoughtful	Clever	Wise
Management Guru	Peter Drucker	Alfred Chandler	C.K. Prahalad

I must add here that this is yet a preliminary table; many of the concepts being presented are still being explored. To the best of my knowledge they are articulated here for the first time in a business book. In a future book we will improve, develop and explore these concepts in a lot more detail - because they appear worthy of such consideration.

Key principles highlight the differences between the three strategies

So let us examine some of the key elements of this table in more detail. The key principles of these three strategies are wildly different. The mathematical principle of the accretive strategy is simple addition, logical and precise and repeatable every time under every condition: 2 + 2 = 4. As we move to the strategy of synergy - the mathematical equation becomes esoteric and imprecise: 2 + 2 ≈ 5. This requires imagination and belief, and only a few people with experience and intelligence can see the synergies, even in the best of the deals. Of course, after the synergies are realised, nearly everyone claims that they could see it quite clearly beforehand. The mathematical principle of synchronicity is yet more esoteric and imprecise: 2 + 2 makes 22 - but only under certain conditions. You need to go beyond imagination and belief - possibly into the realms of visualisation and faith to be able to see the process as well as the outcomes. Not everyone can do it. As I revise this chapter, the example of Apple after the demise of Steve Jobs is staring us in the face. Not much has changed, but the critical element is now missing - it is questionable if the 5-STAR network of Apple will remain the same in the coming months and years.

That brings us to the key economic principle of these three strategies. Accretive strategies are based on the economic principle of scarcity of resources, as espoused by the traditional economic gurus of free markets - such as Adam Smith. Synergy strategies are fundamentally predicated on the economic principles of prosperity through clever actions as suggested by the likes of John Maynard Keynes and his followers. Not many businesses, societies or economies transcend to the next level of economic principles - of an abundance of outcomes based on wisdom, creativity and co-operative effort - that underlie the strategies of synchronicity. The phenomenon is no more imprecise than any other that the economists study, yet they are totally incapable of its study because the language, the tools and the equation of the science of economics are inadequate for its study. Imagine an economist trying to explain the success of Apple under Steve Jobs. You could equally try and imagine a man from 200 years ago trying to explain the functioning of radio, electricity or television.

Metaphorically speaking, the three strategies are three levels on a Jacob's ladder

Metaphors are sometimes useful to convey an idea - especially a new one. While they are not to be taken literally, they do accentuate the differences between the three strategies markedly. A physical metaphor that makes the difference clearer is that accretive strategies are similar to Newtonian physics, while synergies are like quantum physics and synchronicity is akin to Einstein's theory of relativity. For the non-scientists, a visual metaphor might make the picture a lot clearer - imagine the difference between a torch light and a laser beam - that is the difference between the accretive and synergistic strategies.

Now imagine the difference between the laser beam and sunlight where the intensitiy of the sun's light is present all around - that is the difference between the synergy and synchronicity.

The economic metaphor for the accretive strategies is free trade and the management guru who follows this logical path most precisely is Peter Drucker. The economic metaphor for the synergistic strategies is the managed market model prevalent today in most countries (even the USA, the erstwhile paragon of free market virtue), and the management guru who fits in with this paradigm most is Alfred Chandler. Finally, the economic metaphor that fits in with the synchronicity strategy the most, is free networks, but it is still not clear if there is a management guru who fits the bill. I would imagine that C.K. Prahalad would have probably have done so, if he was not hampered by the legacies of the past.

Finally, let us examine an organisational metaphor - a well-oiled Swiss chronometer would serve as a good metaphor for the accretive strategy, since all parts play their role to produce a predictable outcome. A well-conducted orchestra is a good metaphor for the synergies, because the final outcome is more than the sum of its individual parts. A suitable metaphor for the synchronicity is a school of fish, or a flock of birds moving in unison - forming a myriad of shapes every few seconds without any central control.

The journey from free markets to networks is not a straight line

One last point before we conclude this chapter, is to mention

Figure 4.2: The path from a free market to free networks and beyond

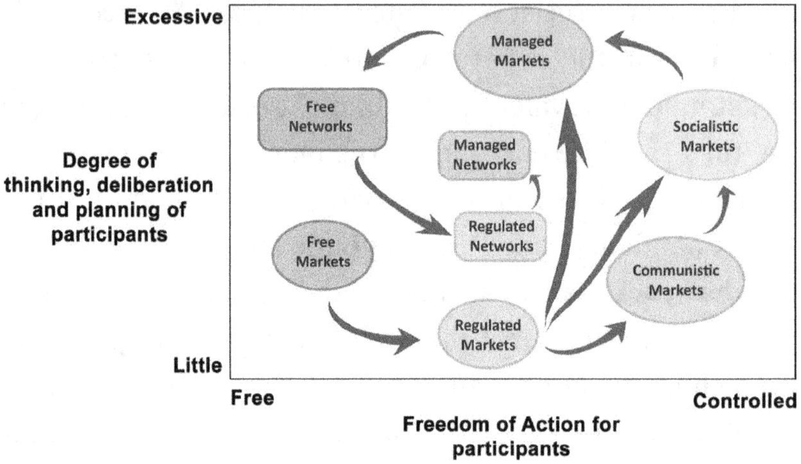

that while Table 4.2 (Value Addition Strategies) gives an impression of a very linear movement from free markets to managed markets to networks, from a still broader perspective it appears that the movement is far more in the form of a "swirlogram" as shown in Figure 4.2 above. While the economists and political pundits will have a lot more to read into this illustration, as pragmatic business people, we will restrict our focus to discussion of the impact of this trend on the businesses.

Starting from the free markets of 19th century, we have traversed, in modern times, through regulated markets to managed markets either directly, or through socialistic and/

or communistic experiments, depending on where you live and work. Right now, we are in the managed markets phase, but moving toward free networks. A full move to free networks might take another two to three decades, subsequent to which we will start hearing calls for regulating networks, which might last several decades and then lead on to managed networks. In a later publication, we might cover these concepts in a lot more detail, but their key relevance to our discussion is that, as prudent business people, we must anticipate the move towards more regulation of the global business networks, and then towards a more managed network structure in the global economy.

In conclusion, business networks release latent energy of the participants, due to the voluntary nature of participation and participants' own self-interest in the outcomes. Yet it goes beyond the transactional interactions inherent in the free markets. With repeated interactions, this massive latent energy makes the outcomes far more superior than any other form of association.

SUMMARY:

Networks thrive on trust: they succeed where trust building mechanisms are strong and well adhered to. Well functioning business networks incorporate a secret source - the power that goes beyond the synergy, the multiplier effect, into the realms of synchronicity. Besides, business networks allow businesses to be simultaneously strong in their core strengths and live with their own weaknesses - the other members of their business networks make up for their weaknesses. As businesses move from traditional structure to business networks, the essential co-operation building mechanism moves from control to co-ordinate to co-create.

The story of Red Bull - a case example

We will see many examples of networks in action displaying their synchronicity throughout this book. A case in point is the story of the energy drink Red Bull. The tale defies all explanations. There was a time in early 1980s when Red Bull's co-creator and celebrated face - Dietrich Mateschitz - grappled with seemingly insurmountable problems within his entrepreneurial venture. The Thai drink adaption was considered 'disgusting' and 'unappetising'. The market research firm was very clear in its conclusion "No other new product has failed this convincingly". Health authorities were suspicious of the ingredients and were reluctant to grant permissions. On top of it all, Mateschitz, a marketing professional, lacked manufacturing and supply chain experience. His partners in the venture were a father and son team of Thai product developers and businessmen, who lacked any real international savvy.

Three things that Dietrich Mateschitz had in abundance were business savvy, an infectious can-do attitude and an intuitive understanding of the power of business networks. Those allowed him to fight some of the biggest and most valuable soft drink brands on earth and create a niche in which he still reigns. The entire story of the company is told in great detail in a number of books and articles, and generally each writer has explained it through their own lens - somewhat like the parable of five blind men describing an elephant. Here is my own brief, albeit somewhat one-eyed account of the key victories along the way and the role of networks in this success story.

From the very beginning, Mateschitz built a network of loyal fans, customers, marketers, logisticians and distributors. Realising that there was no way he could take on the massive marketing and advertising budgets of the large global beverage marketers, he built unique relationships with his key marketing enablers. For example, marketing guru Alex Wipperfurth in his book "Brand Hijack" said Mateschitz's relationship with his advertising agency went far beyond the customer-supplier relationship seen in the corporate world. "It is probably the strongest client/agency relationship in the business, with the ad agency as true, long-term partner rather than dispensable vendor."

By far the strongest network that Mateschitz created was of loyal customers. Fueled by innuendo, rumour and speculation about the special powers of the product and its ingredients, the early consumers saw themselves as special people participating in a network of counter-cultural movement. That Mateschitz has managed to grow the network to its immense size and keep its spirit alive is a true testimony to his marketing genius and will make a great case study in network creation and nurturing.

Equally strong and loyal networks were created in distribution, sales and logistics. Once a critical mass was created in a particular geography, the distribution network was almost always on an exclusive basis. This allowed the distributors to focus on the unique ingredients of Red Bull's distribution success and be part of the action - which is incidentally the driving philosophy behind Red Bull.

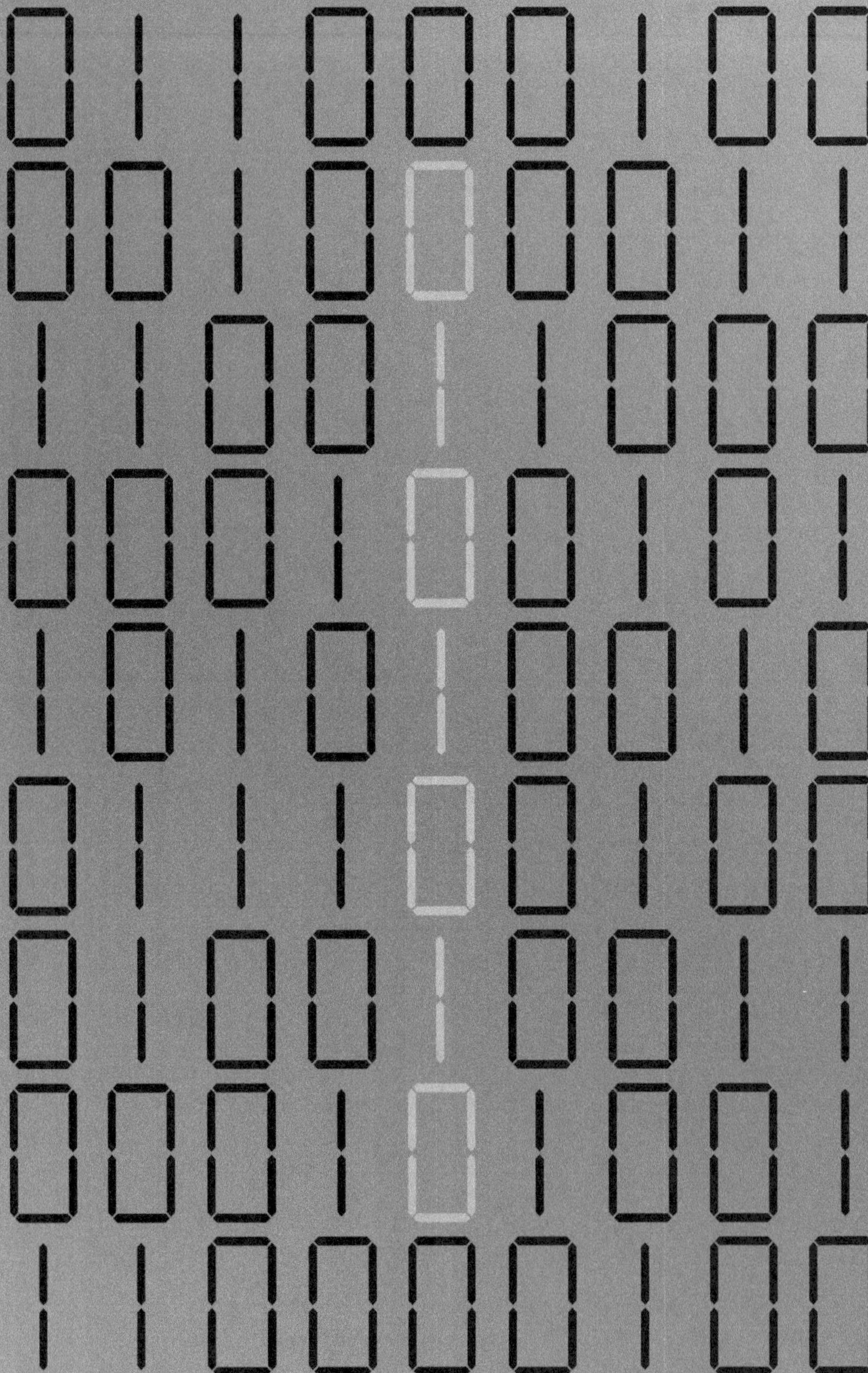

INFORMATION TECHNOLOGY SYSTEMS
SUPPORT GREAT BUSINESS NETWORKS

Chapter 5

Truth about IT Systems Required for Great Business Networks

> ***Anyone can make the simple complicated; Creativity is making the complicated simple.***

Charles Mingus (1922-1979)

I see a lot of business executives switch their minds off when the talk turns to IT systems - so we will exercise a lot of creativity in this chapter to make the discussion as simple as possible. However, this message is important, and we will keep the discussion mostly business focused to hold your interest.

There is now a rich legacy of systems and knowledge about them that enables smarter decision making, by understanding the problems that surfaced in systems renewal efforts. We will focus on these problems in the first part of this chapter. In the second part, in order to understand the state-of-the-art systems, we will take a very brief journey in their evolution to their current state.

Having worked for a software company that developed

and sold the kinds of systems that underpin supplier network management, and having developed and marketed such systems, I have some understanding of the associated history and trends, as well as what makes a good system. In fact, I have also written a fairly technical book titled "Sink or Swim Report - How Information Technology can Save or Ruin Supply Chains" (http://bit.ly/ITSink)[1] on the topic, which is meant only for serious practitioners in that field. This chapter presents only a very simplified and abridged version of that work.

Why the discussion is necessary

The first question anybody will ask is "Why is this chapter necessary?" I am writing this rather long chapter because any discussion of the 5-STAR Business Networks without devoting at least some time to the systems that underpin them is like studying the Bible without mentioning Jesus. As you will see in the next section, when we get to more esoteric material about the new model, IT systems underpin many of its components. We will talk about the actual functionality of the systems themselves when we discuss the components of the model; but first, we want to discuss the overall configuration of such systems in the context of historical and current trends.

On one hand, supply network IT systems can yield immense results to companies that implement and use them properly. On the other hand, there are multitudes of stories of very large corporations that have wasted hundreds of millions of dollars on the implementation of such technology systems with the expectation that their business will run far more smoothly, only to find the results are worse than they were before the change.

From the IT company's perspective, the world looks very different

How can you make sure that, when you embark on large-scale supply network information technology re-configuration, it will succeed in getting the results you want?

Let us look at the world from the system provider's perspective. The IT systems providers do not set out to dupe corporations. They generally want to help corporations achieve impressive results. However, they operate in a world of their own. They understand, by their very nature, IT systems are hardcoded to do specific work. They can build a certain amount of flexibility into these systems, but this comes at the price of increased complexity.

At the same time, implementation of complexity costs a lot of money. Hence, information technology service providers have to make decisions where they balance flexibility against the envisaged complexity in the system. They come to a point where the system is not too complex and not too expensive to implement for most of the companies in the marketplace.

The first few clients have a big influence on the software system

Most systems are built to correspond and work very closely with the processes and the supply network configuration of their first few customers, who have exercised tremendous influence on the configuration of the system. Going back to the origin of most companies, we discover that most systems are ideal for a particular set of clients who are similar to the first two or three customers of the systems provider.

As more customers are added, the systems become more complex. Finally, the package comes to a point of maximum complexity, beyond which it is impossible to provide a standard system that can be any more intricate. After that point, the system is marketed as a package that needs to be customised for each new customer.

Many large-scale information technology deployments derail!

Data, anecdotes and case histories abound on the misapplication of information technologies for supply networks. Not too many years ago, a worldwide corporation made news with the downgrading of its earnings expectations after supply chain system implementation setbacks. The expectation was that the new system would reduce the new production cycle from one month to one week. Furthermore, it would better match the demand and supply of its products to place the correct products in the right locations and quantities, all at the right time - a very lofty goal. The company spent an enormous amount of money (exceeding $400 million) in order to achieve its aim. However, the software system 'never worked right'. It caused the factories to crank out too many unpopular products and not enough of the trendier ones in high demand. While making the earnings downgrade, the CEO asked the rhetorical question: "Is this what we get for $400 million?"

The analysts were not surprised. One respected market analyst [AMR] commented: *"Fiascos like this occur all the time, but are usually kept quiet unless they seriously hurt the bottom line."* Another respected analyst commented that while the CEO made it sound like it was a surprise for him, if he did not have checkpoints for the projects, he did not have control over his company. A third commented that companies are confused by escalating market hype and too often underestimate the complexity and risk involved. A different analyst, Forrester Research, commented *"When the software projects go bad, companies are more likely going to scurry up and cover it up because they fear that they are the only ones having trouble. But far from it; our conversation and research reveals this company was not unique or the only one having this kind of trouble".*

Despite their lofty goals, many of the large information technology deployment projects derail. It takes time for word to filter out because, in most cases, the executives involved in the process are far too embarrassed to talk about what happened. They do mutter among themselves and after several similar instances, the mutterings become more vocal and a trend emerges where a number of people start talking about the shortcomings of the system itself, or its implementation process, or the time taken for implementation. Because the cost of failure is so high – greater than $400 million in the above case – it is instructive to understand the real root causes.

30 years of accumulated wisdom is available

It would be a fallacy not to learn from all the accumulated wisdom of the past. After all, those who do not learn from history are condemned to repeat the same mistakes. This will enable us to understand the steps we can take from the very beginning to increase the probability of success. This will also allow you to confidently move forward with business network IT system selection, integration and use to obtain the results you set out to achieve.

The supply networks information technology projects have become larger and larger over the past 15 years. It is quite customary to start with an expectation of spending $50 million, and end up paying in excess of $200 million on systems renewal projects.

Rough estimates indicate that, even today, about one-third of these projects are cancelled without delivering any benefits, after spending more than $100 million. Another one-third of the projects are not cancelled, but fail to deliver significant parts of what they set out to achieve. Only one-third of the projects achieve most of their strategic goals, but many still incur several budget upgrades and time overruns.

Why is this pattern of failure repeated over and over again?

To answer that key question, let's first examine a typical project cost structure. It is estimated that the software costs are no more than 15-20 percent of the overall cost of systems renewal. Programming and configuration costs run from 20-25 percent. External consulting costs generally associated with process changes run from 15-20 percent, data conversion costs run about 10-15 percent, training costs run from 10-15 percent, system startup costs run about 10 percent, applications support costs are between 5-10 percent and hardware costs are between 2-5 percent of the overall cost structure. Out of these costs, only the software expense generally remains fixed through the systems renewal cycle. Most of the cost buckets are estimated ambitiously at the start and tend to overrun quite considerably as the project progresses.

Programming and configuration costs tend to have the highest cost overruns from 30 percent to as much as 300 percent. Process improvement costs also tend to have high overruns, from 30 to 120 percent. The rest of the costs do not escalate as much, but they still tend to overrun anywhere between 20-100 percent. Generally speaking, companies tend to underestimate quite significantly the task. At the same time, they also tend to quite dramatically overestimate their capability to go through the change, either by themselves or with the help of their selected implementation partners.

The expectations from IT are high

The expectations surrounding the systems renewal are very high. Companies want to leapfrog their competitors by streamlining their business processes, and configure planning, and control systems, to manage their supply networks efficiently. The costs of systems are extraordinary, running into multi-millions of dollars. It is indeed a daunting task, but many companies embark on the journey with the same staff, the same systems vendors, and the same implementation partners, who in many cases, are responsible for their current mess or many other messes.

Rarely do they pause to think about how they will be able to manage such a massive transformation without the help of world's best content experts, who are free of personal agendas or conflicts of interest in the process.

The mistakes in the supply network information technology deployment are encountered during all four steps: IT outsourcing strategy development, IT software and service provider selection, IT service provider implementation and integration and; IT service provider relationship management and service usage. Let us discuss them step by step.

Key mistakes in large-scale information technology deployment

A wise colleague, David Ball, once taught me: *"People don't care how much you know, unless they know how much you care"*. I am listing and highlighting all these mistakes with the intention of building a common, uniform and elevated knowledge base among the senior executives in the user base, in the supply network IT systems companies and in the implementation companies, so these mistakes do not continue to happen in different guises.

Some of the key questions companies must ask even before starting on this journey are:

- Why should they do the systems renewal at this point?

- What are the various options open to them?
- Who will help them in making sure that the internal business people, the internal information technology people, the external information technology service providers, the external process service providers, and the external software providers will all work to the same agenda to achieve the business strategy and the supply chain strategy of the company?

In our view, these are the mission-critical questions. Delegating the task of answering these questions to a part-time team of internal executives who concurrently have demanding business responsibilities and attend occasional steering committees meetings would consign the system's renewal efforts to failure from the start.

For ease of discussion we have divided these mistakes into four categories as shown in Figure 5.1 on the next page.

Information technology outsourcing strategy development is rarely thought through

The most common mistakes are improper ROI calculations and inadequate budgeting for the cost and time required to achieve the results. Nearly 84 percent of companies reported after the fact that they should have allowed for more time and more money.

Often, there is a major disconnect between the business strategy and the capability of the IT system that is being bought.

Why does this happen?

Figure 5.1: Key mistakes in supply network IT systems deployment

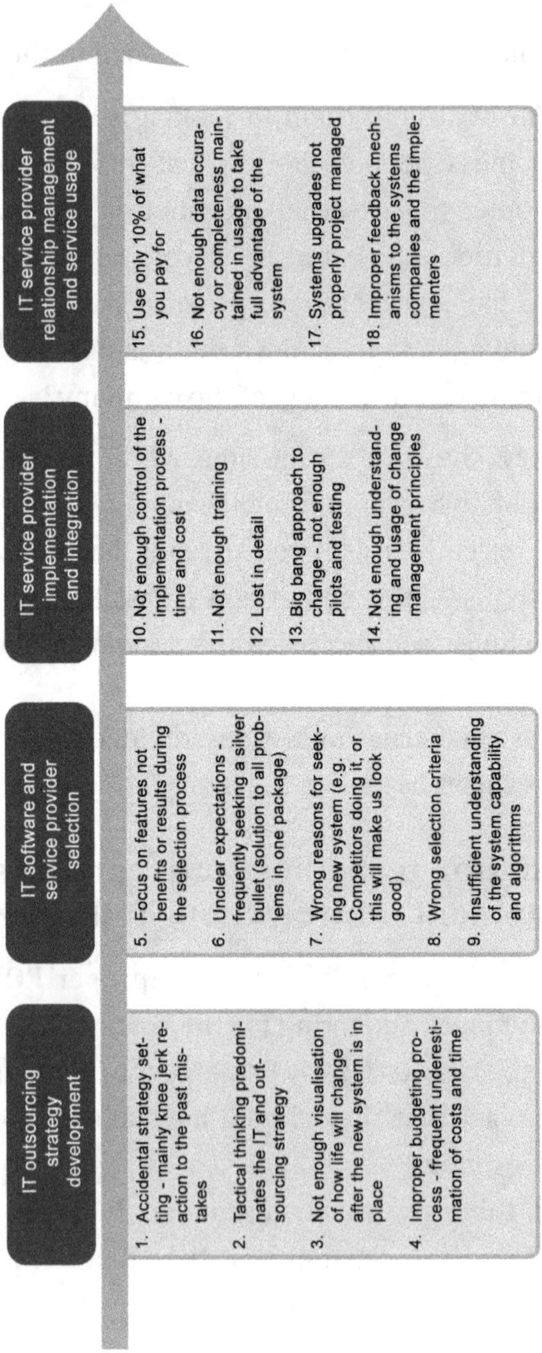

The three different strategies – business, supply chain and IT - need to resonate with each other. For example, if the business strategy is to serve premium customers with customised bundles of products and services at a premium price, then the supply chain strategy is towards the 'make to order' segment of the continuum rather than the 'make to stock' element. A 'make to order' supply chain strategy necessitavtes a much speedier information technology strategy, so systems can respond flexibly to both the supply chain strategy and the business strategy.

On the other hand, if the company hopes to achieve high volume through marketing low-profit, low-cost products to mass markets, a 'make to stock' supply chain strategy makes a lot more sense. Supply chain IT systems will tend to be more rigid, more efficient and oriented around forecasting demand and managing the safety stocks of the business. Responsiveness is not the key for these kinds of systems. However, forecasting ability and management of inventories through the system will be critical. All the decision support capabilities of such systems must be oriented around their ability to safely manage the stock through the ups and downs of seasonality and business cycles.

As the market situation or economic conditions, or the geopolitical backdrop of the business strategy change, these three strategic elements must be constantly updated and clearly articulated to the business network.

In the real world, the three groups frequently act in isolation. The strategists make strategies up in the air, the supply chain executives direct the supply chain down in the warehouses based on their own assumptions about what the business wants to achieve and the information technology staff work away in their cubicles to provide systems based on poorly articulated needs of the business.

In the absence of a clear strategy, tactical thinking takes over

When the strategy is not clearly defined or articulated to the information technology staff, they start making their own assumptions about the wants of the executive team, which sets the strategy. Or, IT start reacting to the needs of the supply chain staff, who themselves are not clear on supply chain strategy or business strategy, but are very vocal in articulating the tactical problems they face in their day to day running of the business.

Though it is important to address these tactical issues in the systems renewal effort, they should be escalated to a strategic level to be addressed. The following example will give you a better idea of how tactical thinking dominates the plans of information technology staff during system renewal efforts.

A recent examination of a business that I worked will revealed that the selection of systems vendor was made entirely on cost constraints, rather than envisaged benefits. As a result, the new system put in place was inferior and did not support the demand forecasting process or the inventory optimisation process of the company.

It is pertinent to point out that this business operated in

an industry that largely thrives or fails on the basis of its inventory management capability. However, the information technology personnel, who made the systems selection decisions, were either unaware or ignored the fact that in the mass retail business in which their company operated, one of the most important determinants of success was the inventory management capabilities.

IT staff were very much aware that the business shareholders and the board were extremely cost conscious and they reacted to that economic constraint far more than to the strategic needs of business for a robust inventory planning system. The entire effort of two years of systems selection and implementation was wasted as a result of this key tactical blunder. Unfortunately, we were called far too late, and it was already way behind its competitors and on the verge of bankruptcy, due to misapplied inventory management capabilities. We could immediately diagnose the key strategic blunder that was made several years ago when the new system was put in place. Penny-pinching at the time of systems renewal had sunk the business beyond redemption.

Lack of visualisation leads to fumbling around

Teams often do not attempt to visualise how the day-to-day life will change once the new system is in place. Without such visualisation, perhaps from inadequate or improper understanding of the coming change, companies are charting a course while surveying the territory. No wonder they get lost.

What do we mean by visualisation of how life will change? The fact is that every company is currently operating under a set of constraints and assumptions. When the new system is fully functional, many of these constraints will disappear and many current assumptions will not be valid. A new way of functioning will need to evolve.

With the availability of knowledge about the new systems, it should be easy to understand the complete capability of what is going to be put in place. With this understanding, the company should be able to create a reasonable view of how the entire business will work. This visualisation allows the company to align the new processes with each other and with the strategy of the business. However, a lack of this vision results in critical mistakes being discovered at every point during the execution. In addition, the staff end up using no more than 40 percent of the functionality that the company enthusiastically paid for.

Budgeting is still 'massaged' to game the decision making

The fourth critical mistake is an improper budgeting process. I have already outlined above how most projects end up over budget and over due. Cost escalations are the norm rather than exception. I believe that this is a fundamental strategic mistake. Vendors have an incentive to give out information strategically and only when they have to. It is an imperative for the company to keep digging for information to enable it to budget fully before the project goes past the comparison of various options stage.

Just because the vendors do not give out enough information to create cost comparison, this is not an excuse for a company not to do their budget in sufficient depth and with adequate rigor.

IT vendor selection process is muddled throughout

Many companies have a democratic systems selection process with a 'one person, one vote system'. However, they forget that it is easy for many staff members to be distracted by flashy system features that will yield hardly any practical benefits to the users.

It is customary to blame the system for process or skills shortcomings. It is equally customary to have an expectation that the IT systems upgrade or new deployment will resolve these shortcomings automatically. In most instances, nothing could be further from the actual outcomes.

There are many legitimate reasons for seeking new IT systems; enhanced capabilities, measurable and clear increases in supply chain efficiency and effectiveness, support for business strategy, and so on. Many companies seek new or upgraded systems for the wrong reasons, such as competitors are doing it or this will make us look good. Coupled with the above three problems, this is the dilemma of unclear or wrong selection criteria.

Frequently, companies fail to look behind the veil and discern that what appears to be a single system is actually a mishmash resulting from multiple acquisitions. Risks are frequently underestimated and benefits are overestimated. Realistic expectations, clearly articulated selection criteria, along with weights attached to each, weighted voting rights, and well-developed systems selection process based on meaningful benefits, are some of the ways to achieve better results.

Most decision support systems have sophisticated algorithms built into the code. All systems have process assumptions built into them. In many cases, the selectors do not adequately understand these algorithms, process assumptions and systems capabilities. The decisions are based on a vague understanding of the capabilities and algorithms. No wonder the results fall short of expectations.

During the systems integration, users frequently lose control of the project

During the implementation process, the flaws in the previous two steps clearly manifest themselves. Time and cost overruns are frequent. Few projects, if any, deliver on time precisely what they set out to deliver. Frequently, companies lose control of the implementation process from the beginning by relying far too much on the implementation partners.

Another common complaint from users is the insufficiency of training. Frequently, due to budget overruns towards the end of the process, training is cut short. All this minimises the benefits of the new system.

By the time the systems implementation process is underway, most people have lost sight of the original strategic objective. Goal posts are changed mid-way through the process by the inability to see the final outcomes clearly.

A 'big bang' approach to change is another key mistake, where enough pilots and testing are not undertaken. This used to be a common problem, but is still encountered sometimes, where companies take on too much change without enough testing.

Similarly, not enough understanding and usage of change management principles is also a major mistake. Despite all the focus on change management over the past decade, critical steps are often missed in the change management process. Stakeholders are not adequately covered, top management is not fully involved in the process, and incentives are not properly aligned for all the stakeholders.

On-going relationship management is frequently formulaic and lacks substance

Because of insistence on flashy features during the systems selection process, and inadequate training, companies frequently end up using only a very small percentage of all the functionality that is paid for. It is also common to pay significant sums for customising the systems with functionality that is later found to be redundant or unusable.

Despite the fact that one of the main reasons for IT systems upgrade or deployment is the lack of data accuracy and completeness, it is frequently discovered after the completion of the project that the original problem remains. There are many reasons for this, mainly encountered in the first two steps. After the initial systems deployment, further upgrades are frequently left up to the mercy of the vendor. A streamlined process is rarely deployed to assess each upgrade for its suitability to the business strategy.

Because of the lack of skills and a rigorous process in the first three steps, it is rarely possible to isolate root causes or most of the problems encountered during the journey. The systems companies and the implementers are rarely held accountable for their respective roles in accentuating the chaos. Inadequate feedback results in problems manifesting themselves over and over, until a sufficient body of disgruntled customers builds up, who gradually air the problem in vague terms. Lacking in timely and accurate feedback, improvement is only gradual and sporadic.

This is where the benefits from the previous three steps are expected to manifest themselves and make it all worthwhile to have made the effort. We frequently find that because of all the other critical errors made during the previous three steps, companies sub-optimise the usability of the system quite dramatically and fail to gain the expected jump in productivity, competitive advantage or enhanced effectiveness in their supply chains.

I make no apologies for taking the time and making the effort to clearly articulate the problems with the IT systems that fail in supporting the business networks. No business network can truly be great (in both its effectiveness and efficiency), unless it is underpinned by a robust and functional IT system.

So what does a functional IT system look like? What would support a business network of the future?

This is the key question we address in the rest of this chapter. Let us see how the business networks have evolved in the past few decades, and how the functional IT systems have evolved in line with the evolution of the business networks.

Evolution and trends in Information Technology Systems underppining complex business networks

To be able to understand the current state of technology, and to project the future trajectory, it is important to briefly understand the evolution and trends over the past five decades that have underpinned complex business networks.

As shown in Figure 5.2 on page 85, we take a brief snapshot of each decade to discern key movements in the information network capability and describe how it has consistently underpinned the business network capability.

Supply network management strategy and capabilities evolved from primitive to esoteric in the four decades since the mid-1970s:

In the initial phase (pre-1950s), there was no need of any strategy for supply chains. In the decade that followed, supply chain strategy focused only on the overall profitability of the business. Supply chain strategy came into real existence only after the Materials Requirements Planning (MRP) revolution of the 1980s. Companies became aware of the growing material scarcity and started optimising resources. Strategic activities were in the hands of a few key people in the organisation. Strategic planning was highly inefficient due to lack of data in a standard format. Planning was done without addressing ground-level problems.

The MRP revolution evolved into Enterprise Resource Planning (ERP) during the 1980s and 1990s. Enterprise-wide supply chains were capable of collectively forecasting the supply and demand levels. Product movement through the supply chain to customers was based on assumptions of demand. It became apparent that integration of core transaction processing was critical to achieving seamless integration of business processes.

The strategy was based on the "PUSH" model. However, inappropriate forecasting tools and the "PUSH" strategy resulted in high levels of inventory and underutilised resources. The system used arbitrary management decisions to resolve demand and supply conflicts. A sales plan was used to drive the manufacturing plan and schedule by manipulating individual component quantities and due dates and any capacity imbalances were resolved by adjusting the due dates.

The 1990s and early 2000s' ERP revolution led to a migration from a PUSH model to a PULL methodology (build-to-demand). Globalisation and a search for cheaper locations led to manufacturing outsourcing. The introduction of Just-in-Time (JIT), Lean manufacturing and Kanban by Japan, and widely adopted in the west, led to flexibility in production.

Finally, businesses started to take an external view with the help of strategic partners and a focus on customers. Customer

Figure 5.2: Seven decades of progress underpins current IT systems

Pre- 1950's	1950's - 1970's	1970's - 1980's	1980's - 1990's	1990's – Early 00's	Early 00's – 2010	2010+
• Traditional economy • No tools, no planning	• Profitability calculations • No optimisation of resources and capital	• Ineffective planning due to lack of standardised information, integrated data and analytical tools • Tools capable of optimising internal processes • Dedicated tools for each function • Tools were not robust. Manual intervention was required	• Predict demand by simulation and analytical forecasting tools • Analytical tools came into existence for forecasting and evaluation • Planning process became more integrated and driven by cross-sectional teams • Tools capable of enterprise-wide optimisation • Capability of accurately forecasting demand levels	• Managing demand and understanding customer value proposition • Asset utilisation • Manage lowest cost and distribution services • Faster and more precise operational response • Clear understanding of customer requirements	• Information sharing and visibility across the supply chain became an important factor • Flexibility and responsiveness are required to be competitive • Strategic logistics and manufacturing outsourcing	• Real time information availability in standard format across the complete supply chain • Automated connection in businesses, partners and key stakeholders to share information electronically • Real time visibility across the system to reduce errors in forecasting • The integration of information and decision support systems in supply chain management • Integrated and centralised supply chain planning

Figure 5.3: Evolution of supply network management strategy and capabilities

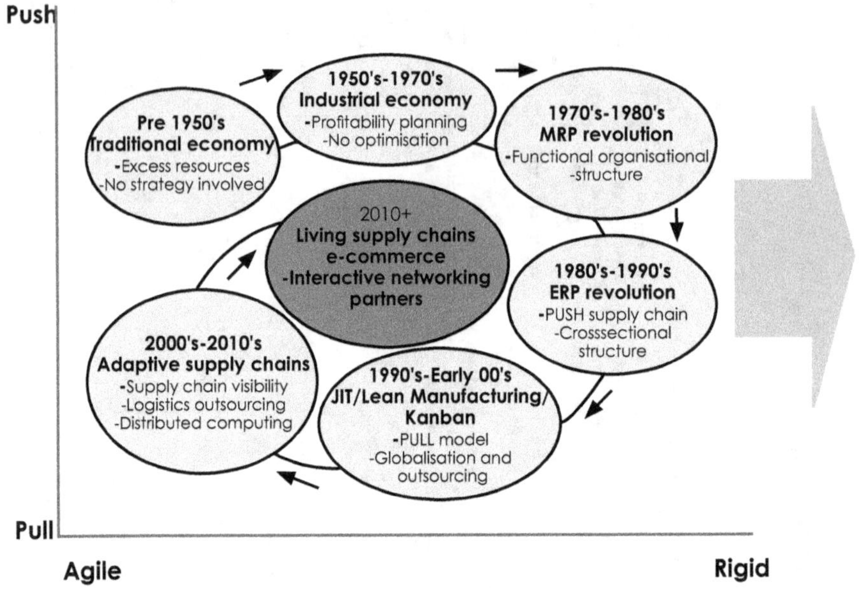

satisfaction became one of the criteria to decide supply chain success, leading to managing total cost and life cycle alliances. Faster and more precise operational response provided capability to meet customer demands (time and reliability).

During the 2000s and into the present decade, trading partners and suppliers collectively developed innovative ways to reach customers. This led to logistics outsourcing and evolution of 4PLs. The supply system adopted organisation alignment towards rapid innovation and there was a quick introduction of new products. Emphasis on rapid decision making and fast response changed supply chains into supply network systems. Focus started shifting towards providing visibility across the supply chain.

For process optimisation, logistics and manufacturing options were strategically decided. Another important trend, that emerged during this period, was the development of trade-offs and balancing of profits, service levels, and capacity for different product features for different customers in different channels. There was a rapid and seamless communication of order information across the chain. Other important developments in supply chain systems were synchronisation of supply chain activity with capacity utilisation, capturing and translating customer demand into line rates, build schedules, and material call-off and logistics.

In the future, supply networks will evolve as a system of mutually interactive components. They will strive to create value through supply network integration, identifying the right supply chain strategy for a specific product and a particular business model. Purchasing, production, and distribution strategies will be modeled for a global business environment. Strategic partnering and outsourcing will also be incorporated in the supply chain operations. Supply systems will now evolve to a supply ecosystem to address the need of information sharing. Supply chains will also evolve as an ecosystem of mutually interactive components. There will be automated interfaces for supply chain partners to share information electronically. Real time visibility across the system will reduce errors in forecasting. Integrated and centralised supply chain planning will become more effective. Enhanced responsiveness in supply systems will be enabled through direct sales to customers, 24/7 access from anywhere, wider product portfolio and information aggregation, customisation, faster time-to-market, flexible pricing, product portfolio and promotions, efficient funds transfer, user-friendly automated processes and lower stock-out levels.

Processes evolved in concert with the strategies

Trends in the evolution of supply network management processes show an inclination towards integration and strategic outsourcing. Initial ad-hoc and amorphous systems inside each functional silo were knitted into a hierarchical top-down model that worked together well – but was very restrictive, prescriptive and inward looking. These were in-turn replaced by flat structures at the turn of the 21st century, when the customer was the focus of all activities. Flat structures are giving way to distributed networks of supply systems that can focus on the customer on one hand, and on the profit maximisation on the other.

Software, information and technical architecture evolution underpinned the development of supply network processes, strategies and capabilities

It is clear from the above discussion that supply networks will evolve as an ecosystem of mutually interactive components in the coming decade. There will be automated interfaces for supply chain partners to share information electronically. There will also be real time visibility across the system to reduce errors in forecasting. Integrated and cen-

Figure 5.4: Evolut on of supply network management process

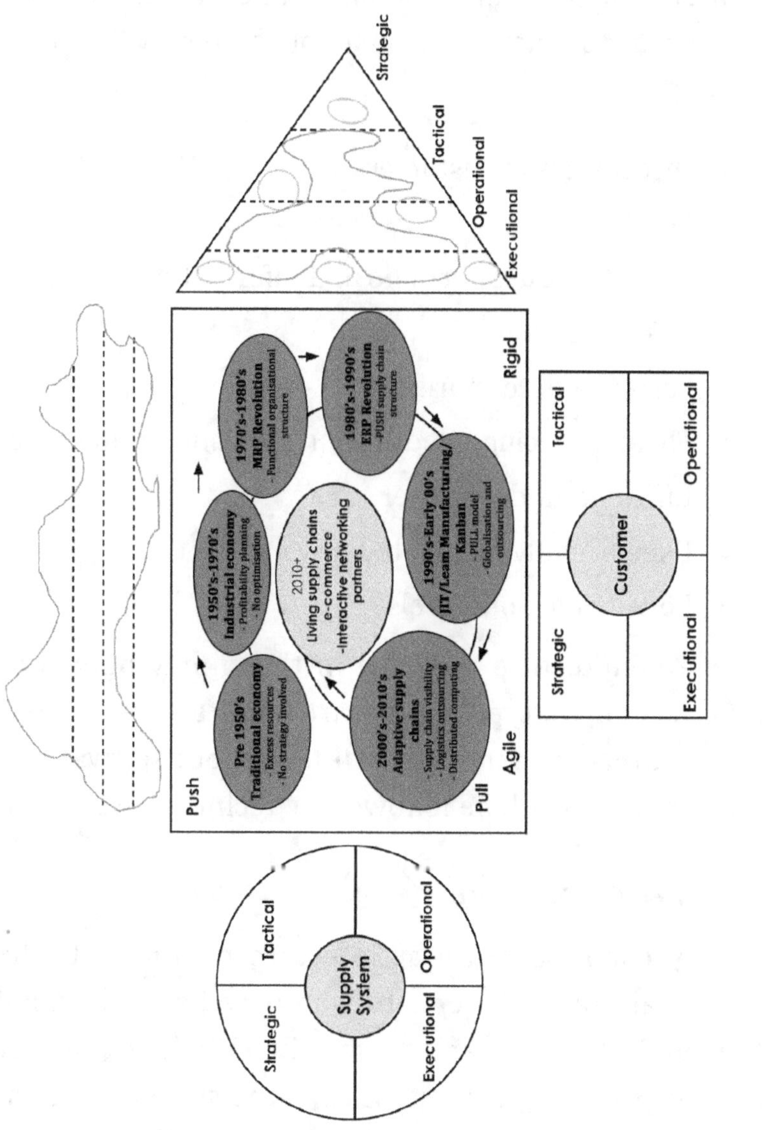

tralised supply chain planning will become more effective. Enhanced responsiveness in supply system will be enabled through:

- Direct sales to customers
- 24/7 access from anywhere
- Wider product portfolio and information aggregation
- Customisation
- Faster time-to-market
- Flexible pricing, product portfolio and promotions
- Efficient funds transfer
- User-friendly automated process
- Lower stock-out levels

The architecture required in the supply network systems are of two types - software architecture and technical architecture. First, we will talk about software architecture, which will be followed by technical architecture.

Software Architecture

Supply chain software architecture continues to evolve from mainframe systems to web-based solutions. Before the 1980s, the software in use suited the functional transactional systems. The activities of the systems were automated and repetitive within functional 'silos'. This led to inconsistent duplication of data throughout the organisation. Also, the systems were 'owned' by the management information services department. In the 1980s and 1990s, systems evolved into enterprise transaction systems with best-in-class tools. The systems were integrated across functions and provided a consistent view of data across the organisation. Unlike in the previous period, systems were owned by the business users and supported by the MIS department. However, the systems were still proprietary with closed architecture.

In the 1990s and early 2000s, best-in-class tools and software emerged, which reduced the role of core transaction systems. Data was distributed throughout the organisation and synchronisation tools assured consistency. New software provided analytical support for strategic and tactical planning.

In the early 2000s-2010s, the software architecture became object-oriented 'plug and play'. The architecture became fully open and software conformed to established standards enabling true plug and play. Instead of depending on MIS, business users started 'assembling' solutions to conform to their specific needs by combining software components.

Technical Architecture

The technical architecture evolved along with the software architecture. Before the 1980s, mainframe systems with 'dumb' terminals were the available technical architecture. Mainframes supported the operational and executional needs of supply chain management, such as inventory management, manufacturing, resource planning and scheduling. These systems were tightly controlled by the MIS 'priesthood'. Information and application requests were often funneled into a 'black hole' backlog. Firms were bound to mainframe vendors by proprietary hardware and operating systems. Vendor leverage led to extremely high costs.

In 1980s and 1990s, minicomputers and personal computers were used along with mainframe computers. Users began to recognise that highly dynamic business trends required more rapid response than was being provided by the current IT environment. Users tried to circumvent the problem by installing departmental or functionally specific systems on minicomputers and/or PCs. While speed of response improved, the 'ad-hoc' approach led to poor maintenance, insufficient training and inconsistent handling of data throughout the firm.

In the 1990s and early 2000s, client/server systems evolved in a big way. Technical advances provided an alternative solution, client/server, for users faced with the problems inherent in a mainframe-based IT architecture. By reducing dependency on proprietary systems, client/server systems shifted power from the vendor to the system buyer. Properly configured client/server systems enforced data integrity by controlled replication and synchronisation and through firm-wide data repositories. Client components provided Graphical User Interfaces (GUI) leading to fewer operator errors and lower training requirements. In the early 2000s, the fully distributed computing model emerged. Business trends demanded that the technical walls between firm, customer and vendor be relaxed or eliminated. Data must move freely across the extended enterprise. The distinction between clients and servers grew ambiguous and peer-to-peer systems became viable. Also, enterprise software was written in object-oriented code to permit rapid development and provide code reusability.

The current state-of-the-art systems are open, collaborative, with easy interface backed by BIG data capabilities

Now that we know the evolution, we should also point out the current state-of-the-art systems. No function, no process and no transaction is conducted today without a supporting system.

On one hand, users prefer light touch capabilities at their fingertips at all hours in the form of iPad or iPhone type apps. Ease of use, speed and, intuitive interface are the key determinants of success.

Hidden behind this simplicity is a very sophisticated capability to collect, collate, cross-reference and analyse a vast amount of data. This creates superior insights that drive the business networks. We will see examples of this capability in Chapters 8 through 12, especially when we discuss the TOP in Chapter 10.

Such capability needs best-of-breed software from the vendors in supply chain decision support, business intelligence and predictive modelling. Obviously, the choice of the right vendor is an art form and we have already discussed the pitfalls to be aware of. Putting together an integrated package of these best-of-breed software is another art form.

The current trend in software architecture is to use internet based supply chain tools. The features of these tools are service oriented architecture (SoA), better graphical user interfaces (GUI), expressive commerce, integrated, interactive software, real-time flow of information (supply chain visibility) and extreme transaction processing (XTP).

The future technical architecture will be enterprise wide web mobile and fit-for-purpose portals. Software solutions will have to address the need for connectivity and security. Software will be capable of handling extreme transaction processing. Adoption of XML seems to have a significant impact on next generation enterprise software technical architecture. Other features of next generation technical architecture are service oriented architecture (SoA) and Software as a service (SaaS).

With emerging borderless networking solutions and communication capability at the fingertips of even remote sites, the entire business network is woven together in an integrated information platform using a variety of fit-for-purpose devices backed-up by big data capability.

There is no doubt that the current state-of-the-art will continue to evolve – the world of information technology never stands still. It is for the executives to choose what suits their current and future business needs and make it work for their businesses and their business network. Those who do so, lay the foundation of success.

CONCLUSION

It is possible to get very technical and details driven when writing a chapter such as this one. However, my endeavour has been to provide as much information as possible in a simple and practical format. For obvious reasons, we have desisted from talking about specific vendors selling software, hardware or networking products. We have also desisted from making any strategic recommendations regarding which approach to take - a one-stop solution vs. a best-of-breed solution. Each of the two approaches had its own merits, though my bias is always towards a best-of-breed solution. But I can understand when a company does not have the time, resources, or intellectual power to engage in a best-of-breed configuration.

SUMMARY

A discussion of the 5-STAR Business Networks without devoting at least some time to the systems that underpin them is like studying the Bible without mentioning Jesus. By their very nature, IT systems are hardcoded to do only certain things. Flexibility can only be built into these systems at the price of increased complexity. A good way to understand great systems is to first look at the common mistakes companies made in their efforts to create IT systems underpinned by their own supply networks. The mistakes in the supply network Information Technology deployment are encountered during all four steps related to its deployment: IT outsourcing strategy development, IT software and service provider selection, IT service provider implementation and integration, and IT service provider relationship management and service usage.

Information technology outsourcing strategy development is rarely well thought out. IT software and service provider selection are muddled through and features driven. During the IT service provider implementation and integration users frequently lose control of the project. In IT service provider relationship management and service usage, there are insufficient ways to measure, manage and improve.

To understand the current state-of-the-art, and to project the future trajectory, it is necessary to understand briefly the evolution and trends that have underpinned the complex business networks. Supply network management strategy and capabilities evolved from primitive to esoteric in the four decades since the mid-1970s. Supply network management processes evolved in conjunction with the strategies and capabilities of the business. Software and information and technical architecture evolution underpinned the development of supply network processes, strategies and capabilities.

The 5-STAR Business Networks

SECTION 2

5-STAR BUSINESS NETWORKS

> ***A designer is an emerging synthesis of artist, inventor, mechanic, objective economist and evolutionary strategist.***
>
> *Buckminster Fuller (1895-1983)*

In the first section of this book we examined the networks in general. We examined what they are, their key characteristics and benefits, their history and the impact of technology.

In this section we will look at the business networks in their full glory. We will examine how modern supply networks are configured for global leadership and how they enhance the profitability of the companies that participate in these networks. We will examine the key cornerstones of the 5-STAR Business Networks and how they underpin success in modern commerce.

Chapter 6 - What Super Networked Businesses of Today Can Teach

- Business model has changed and most people are yet to understand the profound change.
- Traditional organisational chart is redundant. Most companies act as a spear rather a spade in the marketplace.
- The way businesses relate to each other is also changing.
- Traditional camel train model of supply chain is now replaced by a team huddle model of supply network.
- In the 90s the businesses successfully unbundled the corporations into three constituent core competencies.
- Now companies are forming strong business networks that allow them to perform superlatively - efficiently and effectively.

Chapter 7 - Five Key Cornerstones of The 5-STAR Business Network

- Five key cornerstones of a super networked business are:
    - Fire-Aim-Ready (FAR) Innovation
    - $eed-to-$tore (t) Efficiency

  - Transaction Optimisation Profitability (TOP)
  - Advanced Product Phasing (APP)
  - Results-oriented Outsourcing and Modularisation (ROM)
- Excellence in two or three will make any business an industry leader.
- Very few companies do all five at world's best level.
- While a number of companies could be used as a case example to illustrate the concept of 5-STAR Business Networks, Amazon almost excels in all of them.

Chapter 8 - Why Innovation is Not a Solo Sport

- Innovation, besides marketing, is traditionally reputed to be the key engine of growth.
- Product innovation involves newer, better products and services; process innovation involves newer, better ways of making or delivering products and services.
- As companies, products or services innovate, they go through a known life cycle - and business network capabilities must be matched to the requirements on the life cycle stage in order to maximise effectiveness.
- Winning innovators stand the traditional model of innovation on its head by launching products into markets and learning from the customer feedback - the latest example of this is Kindle Serials which allows authors to write books one chapter at a time while taking customer feed-

back for the previously released chapters. I call this model of innovation FAR innovation.

- Supply and business networks must correspond to the market realities and innovation prowess of the company.

Chapter 9 - Cash Rules

- \$t\$ Efficiency pertains to management of cashflow through the business network in such a way that it allows the entire business network to thrive in a sustainable manner.
- Dell is one of the companies which is regarded very highly in management of its C2C cycle which is a vital component of the \$t\$ Efficiency.
- Dell revolutionised the PC industry by lowering the C2C cycle from the industry norm of 56 days to a negative number. Most other super networked businesses also utilise their 5-STAR Business Networks to stand the industry dogma on its head.
- In the retail industry, Amazon has achieved far more efficiency than its competitors, this allows it to offer lower prices and attract new customers achieving extraordinary growth of 100 percent or \$24 Billion in 3 years to 2011.
- A comprehensive framework is required in order to achieve \$t\$ Efficiency. In addition, there are numerous tools such as JIT, TQM, SCOR, etc. which can be deployed

on selective basis.

Chapter 10 - Co-Multiply Your Profits - Use the Full Power of the Global Business Networks

- Imagine a clever shopkeeper who always knows how much a customer would be willing to pay for each item they want to purchase, while at the same time knowing how much each item cost with all the overheads included, as well as how long the item stays in his shop.
- This imaginary shopkeeper will always maximise his profitability for each transaction because he will make all the cost decisions to minimise costs while selling the products at the highest price possible at all times.
- TOP moves towards achieving similar results utilising the power of information technology and business networks.
- The path towards TOP has been paved by activity based costing and balanced scorecards.
- Cost-to-serve modeling is a cruder form of TOP.
- Refinement of cost-to-serve models combined with yield maximisation tools will ultimately lead to TOP.

Chapter 11 - Do Not Be a One Hit Wonder

- Success comes only when the innovation cycle continues in perpetual motion.
- Effort is focused on the current product as well as future products.
- Supply chains change between adaptive and integrative as companies switch from one generation to next.
- It is a balancing act - between the current needs for profits and future needs for products.
- It is unprofitable to launch products too frequently.
- It is suicidal to milk products beyond their useful life.
- 5-STAR Business Networks help take product phasing to a new level.
- Intel has been doing APP very successfully for a very long time.
- From the beginning Intel knew the game was about managing product development cycles.
- Intel also knew how to protect its turf.
- Sales volume feeds ability to invest in product development which feeds more sales.
- Intel is also continually refining its market understanding.
- Supply Chain plays a crucial role at Intel.
- Intel works with its 5-STAR Business Network of collaborators to enable it to launch newer products.

Chapter 12 - Outsmart, Outsource, Outprofit

- Outsourcing has a long history and outsourcing is a fact of life in business today.
- Outsourcing tasks does not transfer responsibility for those tasks.
- Janitorial services are the most commonly outsourced.
- Creative strategists keep coming up with more and more services that can be outsourced so the trend to outsource continues to accelerate.
- Despite its long history and huge growth, outsourcing continues to provide unacceptable results.
- It appears that complexity or size of task increases the probability of failure of outsourcing.
- Two key approaches are being tried to fix the problem:
  - One approach relies primarily on micro management.
  - Euro-centric approach is macro results driven.
- Frequently there is a conflict of interest between the service provider and the customer.
- 5-STAR Business Networks use ROM very effectively.
- By trial and error many businesses have moved towards modularisation of their entire networks but systematic approach to modularisation and outsourcing yield better results faster.
- Modularisation is accompanied by standardisation and

homogenisation.

- ROM puts responsibility where it belongs and makes sure that the customer retains control of the relationship.
- ROM is an intricate dance, which commences well ahead of the actual outsourcing.

SUPER NETWORKED BUSINESSES

Chapter 6

What Super Networked Businesses of Today Can Teach

> ***You never change things by fighting the existing reality. To change something, build a new model that makes the existing model obsolete.***
>
> *Buckminster Fuller (1895-1983)*

I was shocked when I heard of the bankruptcy of Hostess Brands Inc. The $2.5 billion maker of Twinkles and Wonder Bread had about 33 plants, 565 distribution centers and 570 bakery outlet stores across the United States. I have no doubt tons of e-ink will be spilt on dissecting the saga and pinning the blame on the unions, management, workers, government. However, from my perspective, the business model of the company was *passé* long time ago. In this chapter, we contrast the Hostess Brands business model with that of the top-rung players in the food industry - the so-called ABCD: Archer Daniels, Bunge, Cargill and Dreyfuss. If you skim parts of this book, I urge you to read carefully, think and use the material in this chapter because it could make the difference between your company becoming the Hostess of your industry, or one of the ABCDs.

Business models have changed and most people are yet to understand the profound shift

It is a well-known aphorism in the circles of architecture that ***form follows function***. In other words, the structure of a particular unit evolves to facilitate the functioning of that unit. If the form does not match the function, the structure or the function eventually changes.

While researching this book we examined the organisation structure of more than 50 companies, almost all of them looked like variations of the generic structure given in Figure 6.1:

Figure 6.1: Traditional Organisation Structure

Titles and structure in organisational structure vary enormemously. However, most companies agree this is the best way to look at how they are organised to serve their customers' needs.

At the same time, most of these companies evolved at least two decades ago so their functioning has become almost entirely customer-centric, with customers' priorities driving most of the business workings.

In our work with corporations, we have often found the supply network structure frequently results in limiting the effectiveness of the organisation. The traditional structure of the organisation shown above frequently stifles customer responsiveness and innovation. In the modern outsourced globalised world a traditional structure with strict hierarchies and internal walls between departments is a hindrance, rather than an aid for achieving success in business.

In today's hypercompetitive world of consumer electronics, Apple is a standout among peers as great as Sony, Samsung, Panasonic, Motorola, Nokia and Dell. With convergence, all of these companies are vying for their share of the consumer's wallet with increasingly similar products doing similar things. What did Apple do that its market value now surpasses Microsoft's? How did it manage to get over the entrenched competitive advantage of companies such as Sony, Motorola or Nokia?

Your organisation chart is the wrong model for your business

Our work in global supply network strategies and supply network design has convinced us that a modern distributed organisation needs to look at redesigning its structure to keep up with the realities of the business world, which has changed tremendously in the past 20 years. A typical supply chain now runs seamlessly across multiple continents, through boundaries of several organisations, to finally serve a customer with a unique product. To do so, organisations have created de facto structures, which are far different from the traditional structures that they have put in their organisation charts.

The purists might argue that this discussion does not really matter if organisations are already acting in accordance with a de facto structure - an argument which holds some merit. However, people who make up an organisation respond to the explicitly shown structure with much more enthusiasm and clarity than to a defacto or mutually understood structure. I believe that organisations should formalise their defacto structures and use them to gain further competitive advantage. For this purpose we have created the model shown in Figure 6.2 on the next page.

This customer-centric model starts with the customer at the apex of the organisation. Clearly, it is the customer's need the organisation is trying to serve and aligned to the customer is the sales team, who is in direct contact with the customer all the time. The function of the sales team is to have intimate understanding of the customer's needs, customer usage of its products and the customer demographic and psychographic profile. Only then can an organisation create successful products that will gain wider acceptance in its customer base.

Sales are so fundamentally part of an organisation's struc-

Figure 6.2: Customer-centric organisation structure

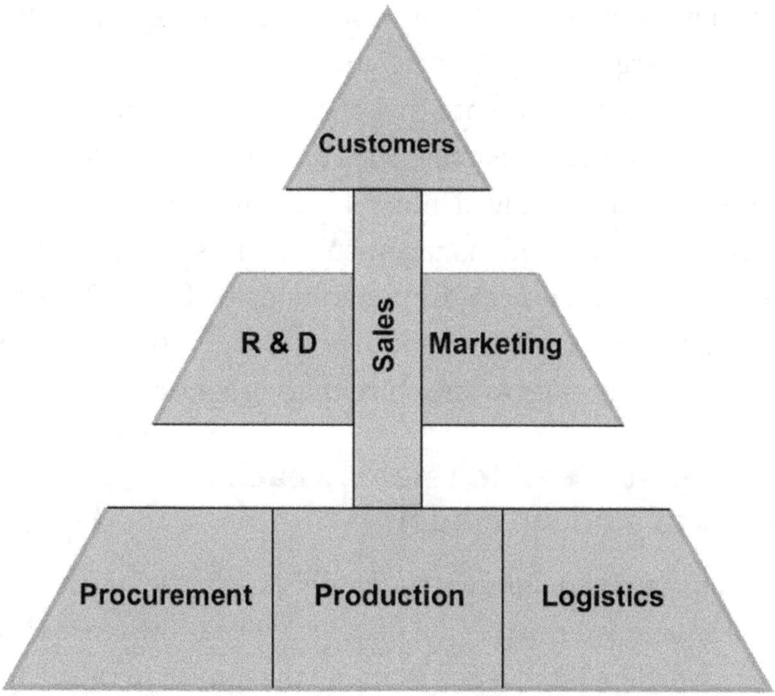

ture that virtually everything excluding sales could be done outside the organisation. Two further key functions which support the sales teams are marketing and research and development. These three form the top-tier of the modern organisation's structure. However, both research and development and marketing, can be outsourced so long as the company and its core team control the outsourced entities.

Forming the second tier is the foundation of the organisation – the supply chain, which incorporates procurement, production and logistics. The supply chain forms the support base or the backbone of the organisation, and is frequently outsourced either to a single provider or to a multitude of best-of-breed providers in such a way that customer needs are met seamlessly without visibility of where any of these activities are actually carried out.

It is interesting to note that this customer-centric model merely illustrates the actual structure that most modern organisations have evolved into. What is surprising is that most business schools and management theories are still persisting with outdated organisational models of the 1980s and 1990s, which bear no resemblance to the actual way businesses are choosing to structure their organisations.

Businesses used to relate to each other in a linear model

The above customer-centric model is a model of the past decade, and later in this chapter, we will see the reasons for this assertion. First, let us examine the impact of this model in practice of the commerce as conducted by many companies today. Due to persistence of traditional supplier-buyer relationships, when this model is applied across multiple organisations it morphs into an unworkable hierarchical structure shown in Figure 6.3 on the next page.

Imagine if five or more organisations are linked in a multilayer structure shown above. Unfortunately, that happens to be the case with many large organisations that compete with Apple in the market-place today. While such a structure minimises cost and responds predictably to all external stimuli, it is not suitable for the world of rapid change we live in today.

Figure 6.3: Traditional supply chain collaboration model

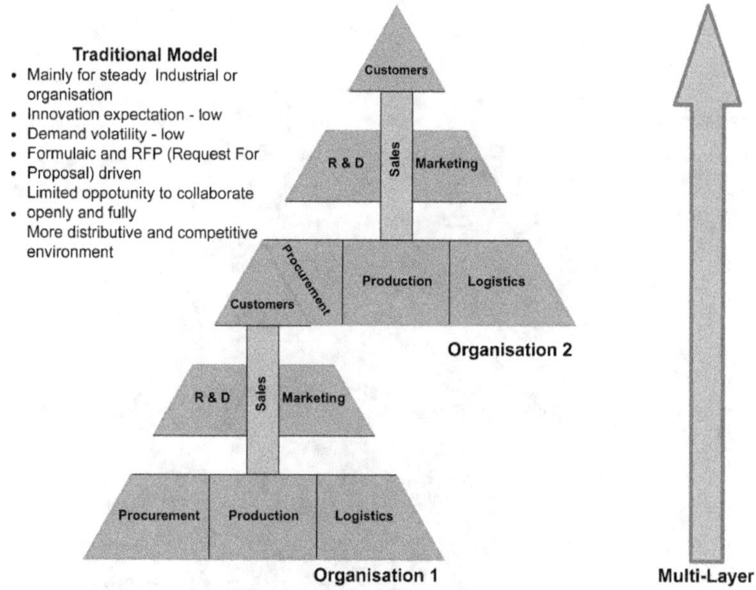

Today, businesses collaborate in a robust network

Apple's success has shown that in the next decade this model needs to be supplemented by an even more evolved model we call the Efficient Global Leadership model (EGL). In this model, we recognise that no single organisation by itself is in a position to service all the needs of a customer relating to even a single product. The fact is that two or more organisations come together as a supply chain and work together collaboratively to fulfill the customer need.

As shown in Figure 6.4 on the next page, each one of these organisations work in close harmony with each other. To create and manufacture products, the produc-

Figure 6.4: Super Networked Business Model

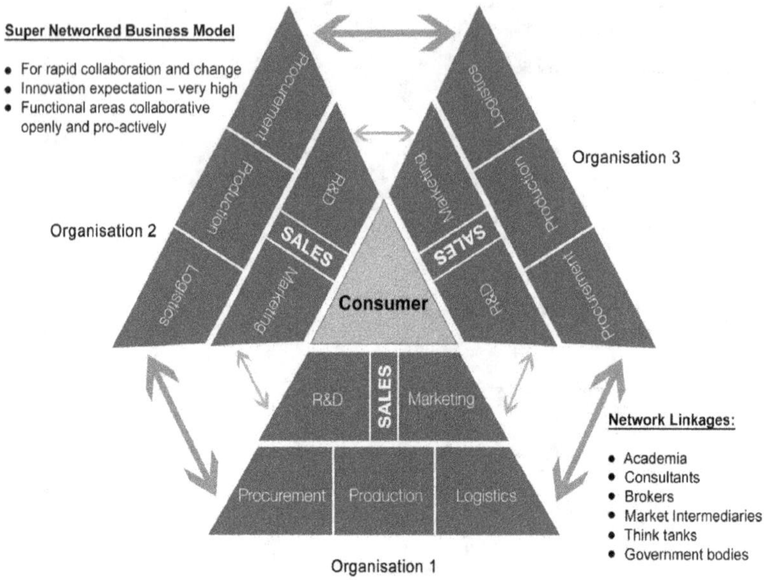

tion teams and the procurement teams work together to put those products in customer's hands. In such a model, a close collaboration is required among the supply chain partners to create, market and sell the products. Similarly, close cooperation is also required to make the products, move and store them in such a way that they are produced in the shortest period of time at a fraction of the cost of traditional products and in the customer hands extremely quickly. Needless to say, when Apple manages to put out one innovative product after another, it is not only its own innovation but also innovation of all its partners at play. Only when companies work together in such an efficient

leadership model do they achieve the level of success Apple has achieved over the past ten years. Figure 6.4 reminds one of the team huddles as shown in the pictures above.

When an individual works on his own, he is neither very efficient nor very effective. That is the key reason, from early civilisation onward that humans have created organisations that give them the benefit of either effectiveness, efficiency or a bit of both. Figure 6.3 (page 115), on the other hand reminds one of camel trains or dog sleds - where one animal is closely following another as shown in the other picture above. Now imagine what would happen to the whole camel train if the first beast lost its way!

Naturally, the question is, why is this important? Think about it for a minute. In fact, put the book down and just reflect on the metaphors. A camel train was a great technology - but is now largely redundant. Moreover, with limited room to collaborate, it is essentially a command and control organisation. In periods of rapid development, if such organisations stick to the tried and tested, they get left behind by their more innovative peers.

Organisations were already unbundling into their core competence

In a highly recommended and seminal article in the Harvard Business Review (March 1999) - John Hagel III and Marc Singer argued that each organisation in fact fulfills three functions simultaneously. Underneath the hood most companies comprise three kinds of distinct businesses: customer relationship management, product innovation and infrastructure management. Because the authors argue their points a lot more succintly, here is an abstract of the article:

> *No matter how monolithic they may seem, most companies are really engaged in three kinds of businesses. One business attracts customers. Another develops products. The third oversees operations. Although organisationally intertwined, these businesses have conflicting characteristics. It takes a big investment to find and develop a relationship with a customer, so profitability hinges on achieving economies of scope. But speed, not*

> *scope, drives the economics of product innovation. And the high fixed costs of capital-intensive infrastructure businesses require economies of scale. Scope, speed, and scale can't be optimised simultaneously, so trade-offs have to be made when the three businesses are bundled into one corporation. Historically, they have been bundled because the interaction costs--the friction--incurred by separating them were too high. But, we are on the verge of a worldwide reduction in interaction costs, the authors contend, as electronic networks drive down the costs of communicating and of exchanging data. Activities that companies have always believed were central to their businesses will suddenly be offered by new, specialized competitors that won't have to make trade-offs. Ultimately, the authors predict, traditional businesses will unbundle and then rebundle into large infrastructure and customer-relationship businesses and small, nimble product innovation companies. And executives in many industries will be forced to ask the most basic question about their companies: What business are we really in? Their answer will determine their fate in an increasingly frictionless economy.*

I believe the authors' prophecies from nearly 15 years ago are now on the verge of coming true - albeit not entirely in the way they foresaw. Most organisations are already showing signs of becoming predominantly proficient in one of the three arenas - customer relationship management, infrastructure and assets management or competency and innovation management.

As the organisations have focused on their core competencies, they have become specialised in infrastructure or innovation or customer service. This unbundling of the corporation is shown as Phase I in Figure 6.6 on page 122, and was predicted in the article by Hagel and Singer. As a result of this unbundling, some corporations - those who focused on infrastructure became highly specialised in asset management. They measure efficiencies in fractions of time and money and are always striving to be more efficient. Toyota's never-ending quest to lower the change over time comes to mind when discussing these types of organisations. Most

Figure 6.5: Unbundling of the Corporations as predicted by Hagel and Singer

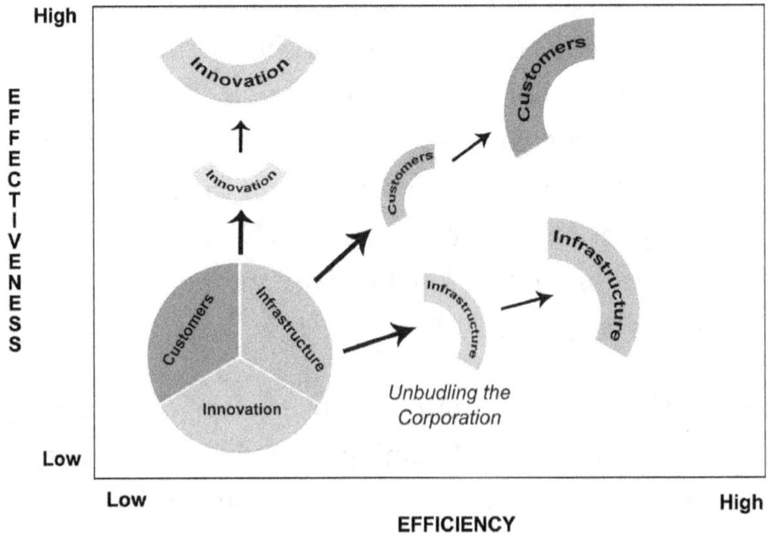

shipping companies and airlines are examples of infrastructure based businesses.

On the other hand, those organisations that focus on innovation and competencies became highly specialised in innovation and new product development. These rated very highly on effectiveness but are not very efficient in their operations. Mercedes Benz (prior to its purchase of Chrysler) is perhaps the best example of a large corporation that takes this route. Most investment banks, consultancies and other competency-based businesses also sit in this camp.

In the middle sit companies which are very profitable - due to their closeness to the consumer and deep understanding f the consumer psyche - but are neither very efficient or very innovative. Travel agencies and freight forwarders are typical examples of these type of companies.

5-STAR Business Networks allow re-bundling in a way that maximises both efficiency and effectiveness at the same time

Most corporations are now gradually and somewhat unknowingly evolving into Super Networked Businesses as shown in Phase II of Figure 6.6 on the next page. Obviously, some companies are traversing the journey quickly and deliberately, while others are stumbling, muddling and copying their way to the end result. However, most companies have come to an intuitive understanding that they can extend their effectiveness and enhance their efficiency at the same time-by utilising their business and supply networks strategically.

While some companies are quite methodically creating super networked business models shown in Figure 6.4, Apple being a shining example of that, others are getting there by trial and error.

This insight has a significant implication for the way business is done. Aided by technology in the global business environment, businesses can simultaneously improve their efficiency and effectiveness. The reason is very simple. If your business focuses on what it is best at and then collaborates successfully with a network of other businesses that make up for the gaps in your businesses' resource platform, the combined whole will have far better efficiency and effectiveness than any of the individual businesses by themselves. That, after all, is the genius of teamwork. Your business network - aided as it is now with the technology and systematic

Figure 6.6: Re-bundling of the Super Networked Businesses

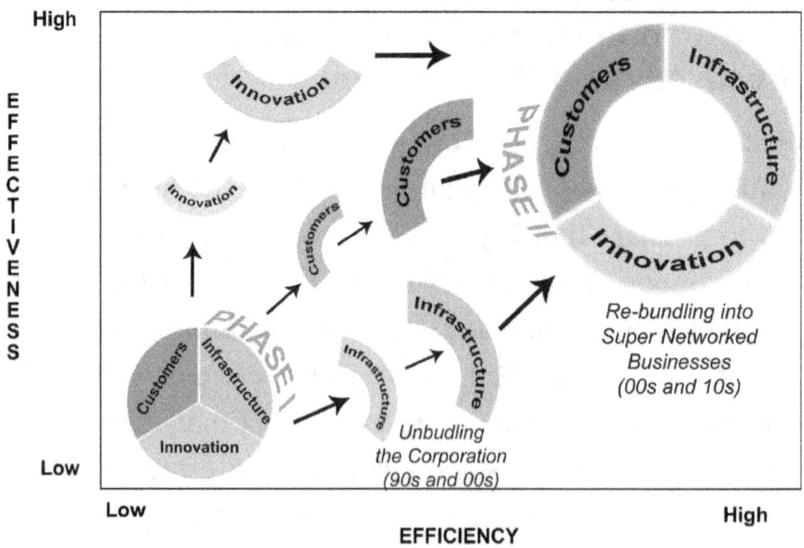

application of methodologies we will discuss in the next few chapters - acts as team.

In the next chapter we will introduce the five key cornerstones of the super networked businesses. These will then be examined in detail in the rest of the chapters in this section.

SUMMARY

The business model has changed, and most people are yet to understand this change. The traditional organisational chart is redundant. Bu necessity, today most companies act as a spear in the marketplace rather than as a spade. The way businesses relate to each other is changing. A team-huddle model of supply network now replaces traditional camel train model of supply chain. In the 1990s, the businesses successfully unbundled the corporations into three constituent core competencies. Now companies are forming strong business networks that allow them to perform superlatively - efficiently and effectively.

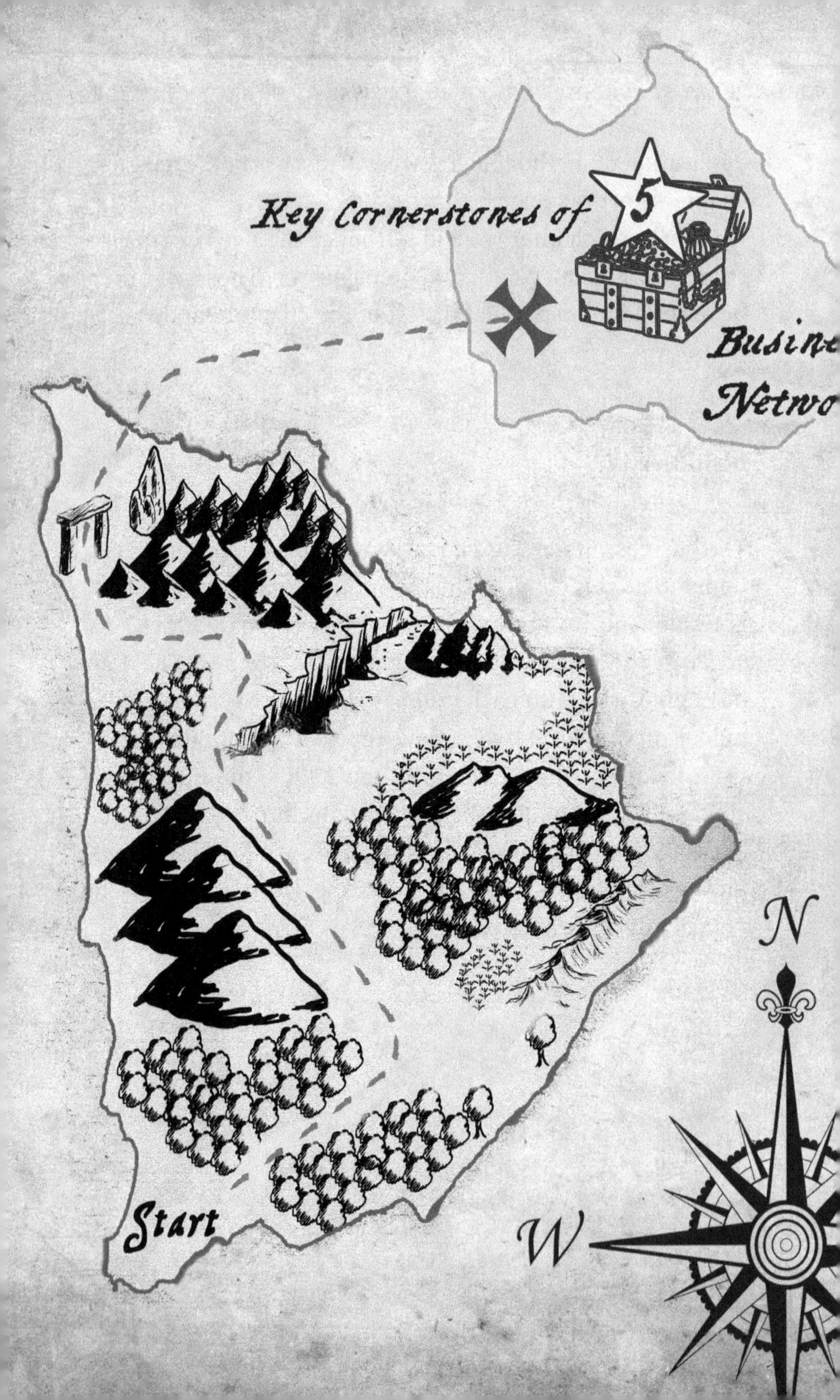
Key Cornerstones of
5
Busine
Netwo
Start
N
W

Chapter 7

Five Key Cornerstones of The 5-STAR Business Network

"We live in a society exquisitely dependent on science and technology, in which hardly anyone knows anything about science and technology."

Carl Sagan (1934-1996)

The office of the Director of National Intelligence in the United States issued its fifth report of a series on Monday, December 10th, 2012. National Intelligence Council's (NIC) latest global trends report, *Global Trends 2030: Alternative Worlds* [1] forms a fascinating read and I encourage you to download it for free from this link, and peruse it at your leisure. The introduction states, "the Global Trends project engages expertise from outside government on factors of such as globalisation, demography and the environment, producing a forward-looking document to aid policymakers in their long-term planning on key issues of worldwide importance."

Networks have become more powerful than they ever were in the past

While the report covers a wide range of issues pertaining to global political and economic long-term consequences, the most relevant excerpts are quoted below (emphasis is mine):

> *The empowerment of individuals and diffusion of power among states and from states to informal networks will have a dramatic impact, largely reversing the historic rise of the West since 1750, restoring Asia's weight in the global economy, and ushering in a new era of "democratization" at the international and domestic level. In addition to individual empowerment and the diffusion of state power, we believe that two other megatrends will shape our world out to 2030: demographic patterns, especially rapid aging; and growing resource demands which, in the cases of food and water, might lead to scarcities. These trends, which are virtually certain, exist today, but during the next 15-20 years they will gain much greater momentum. Underpinning the megatrends are tectonic shifts - critical changes to key features of our global environment that will affect how the world "works".*
>
> *The shift in national power may be overshadowed by an even more fundamental shift in the nature of power. Enabled by communications technologies,*

> *power will shift toward multifaceted and amorphous networks that will form to influence state and global actions. Those countries with some of the strongest fundamentals - GDP, population size, etc. - will not be able to punch their weight unless they also learn to operate in networks and coalitions in a multipolar world.*
>
> *New manufacturing and automation technologies such as additive manufacturing (3D printing) and robotics have the potential to change work patterns in both the developing and developed worlds. In developed countries these technologies will improve productivity, address labor constraints and diminish the need for outsourcing, especially if reducing the length of supply chains brings clear benefits. Nevertheless, such technologies could still have a similar effect as outsourcing: they could make more low- and semi-skilled manufacturing workers in developed economies redundant, exacerbating domestic inequalities. For developing economies, particularly Asian ones, the new technologies will stimulate new manufacturing capabilities and further increase the competitiveness of Asian manufacturers and suppliers.* [1]

This report is nearly 140 pages long and looks in detail at the global trends from a perspective of national governance and security. While the business world is not the prime topic of this report, the way the population goes about conducting its affairs and business has a huge impact on the national governance and power dispersion.

Global business networks will have an increasingly bigger role to play in the future

For this reason, this report creates four scenarios of possible outcomes in the future; I will address only one of the four scenarios - called the Non-State World - which I tend to believe is the most likely final outcome, whether before or after one of the other three scenarios.

> *The present recalls past transition points - such as 1815, 1919, 1945, and 1989 - when the path forward was not clear-cut and the world faced the possibility of different global futures. We have more than enough information to suggest that however rapid change has been over the past couple decades, the rate of change will accelerate in the future. Accordingly, we have created four scenarios that represent distinct pathways for the world out to 2030: Stalled Engines, Fusion, Genie Out-of-the-Bottle, and Nonstate World.* [1]

What is described as the 'Nonstate World' in this report is not a stateless world; rather more of a world with a diminished role of national governments due to individual and business empowerment. With communication, education and sophistication of populations, this type of scenario is indeed inevitable. All it describes is a world where the developing countries will enjoy similar freedoms that the businesses and citizens of the developed countries have enjoyed for decades, if not centuries. The report goes on to describe the networked world in this scenario as follows:

> *Multinational businesses, IT communications firms, international scientists, NGOs and others that are used to cooperating across borders and as part of networks thrive in this hyper-globalised world where expertise, influence, and agility count for more than "weight" or "position."* [1]

Generally, it takes a long time for a trend to register on the radar screen of August institutions. As one of my learned friends is fond of commenting - sitting in the top floor of the ivory tower, the sea always looks calm - it is only when you are battling the waves in a dinghy when you realise the actual situation. Similarly, sitting in a government-funded ivory tower, above comment may be a big revelation, but we already see this reality in the business world around the globe every day.

Successful global business networks will be made up of multitude of disparate players

Indeed, all of the participants named above, the multi-national companies, IT and communications firms, international scientists, professors and consultants and financial institutions have all been working across borders for a long time now. Collaboration, working in a network, and rapid formation and dissolution of coalitions for specific purposes is quite common in this milieu. It is also very true that to thrive in this hyper-globalised world, traits such as expertise, influence and agility prevail rather than "weight" or "position" or any other archaic symbol of power.

In the previous chapter, we discussed the success of super networked businesses and how these networks are coming together gradually. The report referred to above gives us a hint of areas in which to excell for success in the super networked business world. In this chapter, we will examine five key cornerstones of these super networked businesses.

In fact, every super networked business we examined does all these five things very well simultaneously, even though they only excel in one to three of these. Those businesses that excel in all five areas simultaneously are extremely well positioned when compared with their competitors.

There are five key cornerstones of success in building and harnessing effective global business networks

So what are these five key cornerstones of the super networked businesses that lead to these networks being called the 5-STAR Business Networks? As an aid to memory, I have given them mnemonic names in order, shown below:

Before talking about each of the five cornerstones, it must be pointed out that innovation, efficiency, optimisation, product

- **5** • **Fire-Aim-Ready Innovation**
- **S** • **$eed-to-$tore Efficiency**
- **T** • **Transaction Optimisation Profitability**
- **A** • **Advanced Product Phasing**
- **R** • **Results-focused Outsourcing and Modularisation**

phasing or outsourcing are not new paradigms. For example, a study of human civilisation is a study in innovation. Similarly, the Industrial Revolution of the early 20th century primarily focused on increased efficiencies using time and motion studies designed by Taylor and his followers. 5-STAR Business Networks enable businesses to do these regular activities in a much better manner than would have been possible otherwise. As we will see with the help of several examples and cases studies, aided by technologies, an open collaborative mindset and a focus on the bottom line, these businesses are achieving better results through superior methods.

We will examine each of these five cornerstones of the 5-STAR Business Network in great detail in the following five chapters. In this chapter, we will use an example of one company that is positioning itself as the super networked business of this century by using all five of these very astutely.

Amazon is a successful example of an emerging global business network

Figure 7.1: The 5-STAR Business Network

When Amazon was founded in 1994, it was but one of the hundreds or thousands of businesses aspiring to make it big on the Internet. Just like all its peers, initially the markets and analysts were starry-eyed about Amazon's success, and later, when the dot-com bubble burst, and the trend reversed, few people gave it much chance of success. Yet it defied the nay-sayers and continued to sell at PE ratios exceeding 100 on the stock markets.

What is the secret of Amazon's success? What allowed Jeff

Bezos to build the largest online retailer in the world, where customers can acquire anything that they desire over the internet?

Admittedly, the company started with a first-moved advantage in its segment: books. Amazon was one of the first major companies to sell books online. The business was founded in 1994 and by 1995 the website was launched. Initially, the company was exclusively an online bookstore. However, it transformed to sell millions of products to a large and valuable consumer base. Today the company sells everything from electronics to clothing, furniture and even food.

If you had to ask this simple question to 100 people, *"Who on earth today is the world's most customer-centric corporation?"*, Amazon would figure very high on the list. Amazon has achieved low prices, a wide inventory selection, convenience and truly gives customers what they want. As a result, Amazon has evolved into a Fortune 500 business and continues to grow as a world-class electronic commerce platform.

The company grew its annual revenue from US $19 billion in 2008 to US $24.5 billion in 2009 to US $48 billion in 2011, all the while continuing to invest in future businesses and maintaining a healthy cash flow. How does it do this? In an article in Forbes (April 2012) [2] Jeff Bezos offers some tantalizing clues. Bezo's main message is to base his strategy on things that will not change.

For Amazon, their purpose is simple: offer wider selection, lower prices and quick, dependable delivery. Another significant lesson Bezos reveals is obsessing over customers.

Amazon starts with the customer and subsequently works its process backwards. The company even designates specific roles performed by trained employees known as customer experience bar raisers. This is one topic that Bezos takes exceedingly seriously. But, to some extent, every corner store does these things just as well. Why, then, would a corner store owner be lucky to grow his lowly sales by a couple of percentage points, while Amazon grew its sales to $48 billion from $24 billion in just three years.

Amazon achieves high marks on all five key cornerstones of success of global business networks

Let us take a more in-depth look behind the curtains. Jeff Bezos, on the record, said that you have to be willing to be misunderstood for long periods of time. While several of Amazon's designs look like a bust at first, if the new idea makes strategic sense to him, Bezos goes for it knowing full well that people will initially misconstrue the design. In general, this is what innovation is - people are going to misunderstand it because it is new.

Overall, the business philosophy is rather simple: make online shopping simple and suitable so that the customer does not think twice about buying instantly with one click (Anders, George. "Jeff Bezos's Top 10 Leadership Lessons." Forbes. 4 Apr. 2012)[2]. The complexity lies in how this simple business philosophy is translated into consistent action, resulting in nearly a billion customer visits a year. There is nothing simple in the complex execution of this simple business philosophy.

Therein lies the dilemma of the modern business world - the quest for simplicity at the highest level, underpinned by the highest level of sophistication reminiscent of nanotechnology under the hood. Almost all successful businesses do this dance of 5-STAR business network well, but Amazon does it exceptionally well on almost all 5 fronts. There are many other businesses - well-known ones - who could be a poster child for the emerging trend of global business networks we showcase in this book. However, no one is more successful, more visible, has higher potential and is more assured of its role in this revolution. That is why Amazon is a prime example of the 5-STAR Business Networks, demonstrating FAR Innovation, t Efficiency, TOP, APP, and lastly, ROM.

Amazon shines on innovation through its unique approach

"One of the only ways to get out of a tight box is to invent your way out," Bezos declares in an interview [3]. As a whole, innovation is a main area where Amazon shines (Burrows, Peter. "Bezos on Innovation." Bloomberg Businessweek. 17 Apr. 2008) [3]. Preparedness to launch products, services, features and offers that might later fail in the marketplace characterises his approach to innovation. There is no better test of an idea than the customer who faces the moment of truth when they click the buy-it-now button.

Only after exposing it to the harsh realities of the marketplace can an idea be improved by knowing what is missing and how to make it better - an **aiming** process. Kindle is a prime example of this modus operandi. Bezos believes that innovation is being able to spend a large amount of money to construct the capacity that supports new services. Although risky at times, the payoff is greater. In fact, it was Bezos who insisted on the Kindle, which ultimately resulted in a $14 billion profit for the company.

When picking Jeff Bezos as #1 on the Businessmen of Year List for 2012, Fortune magazine dubbed him "The Ultimate Disrupter" [4]. Outlining his achievements the magazine says:

> *Bezos is the ultimate disrupter: He has upended the book industry and displaced electronics merchants. Now Amazon is pushing into everything from couture retailing and feature-film production to iPad-worthy tablet manufacturing. Amazon even sells* **ultracheap database software** *for businesses. (Oracle (***ORCL***), take note.) He's willing to take risks and lose money, yet investors have embraced him, pushing Amazon's stock up 30 percent so far this year. And even as Amazon expands and experiments, Bezos remains zealous about delivering a good customer experience. For all these reasons and more, Fortune has named Bezos its* **2012 Businessperson of the Year***.*

The article covers opinions of a number of industry stalwarts, and I encourage you to read the full article here (http://bit.ly/JBAmazon).

Everything that Amazon does is based on a simple customer-centric proposition: "We innovate by starting with the customer and working backwards," Bezos says. "That becomes the touchstone for how we invent." [4] There is no better example of Fire-Aim-Ready innovation than that statement which stands the traditional Ready-Aim-Fire model of innovation on its head. Not only does Amazon put this into practice, but also facilitates same innovation into its ecosystem of business networks.

How is this done?

Let us focus briefly on the latest service offered by Amazon called Kindle Serials. For a nominal fee (less than $2) a book is delivered to the e-device of the customer at regular intervals. As I write this book I know an enormous period of time transpired between drafting the first chapter and the last. Coupled with a number of re-writes, revisions and edits the cycle time for writing a book can run into years. A lot changes occur within that time frame and early customer feedback would be highly valuable. I know as an author how much I would benefit from a service such as Kindle Serials where engaged customers, having spent less than $2 on a book become intimately involved in the process of co-creation of the book. Kudos to Amazon for making this possible now.

Amazon also scores equally high on efficiency

All that innovation and customer centricity could come at a huge premium. But not at Amazon! The second key underpinning of Amazon's success is the next star in the 5-STAR

Figure 7.2: Fire-Aim-Ready (FAR) Innovation in Amazon

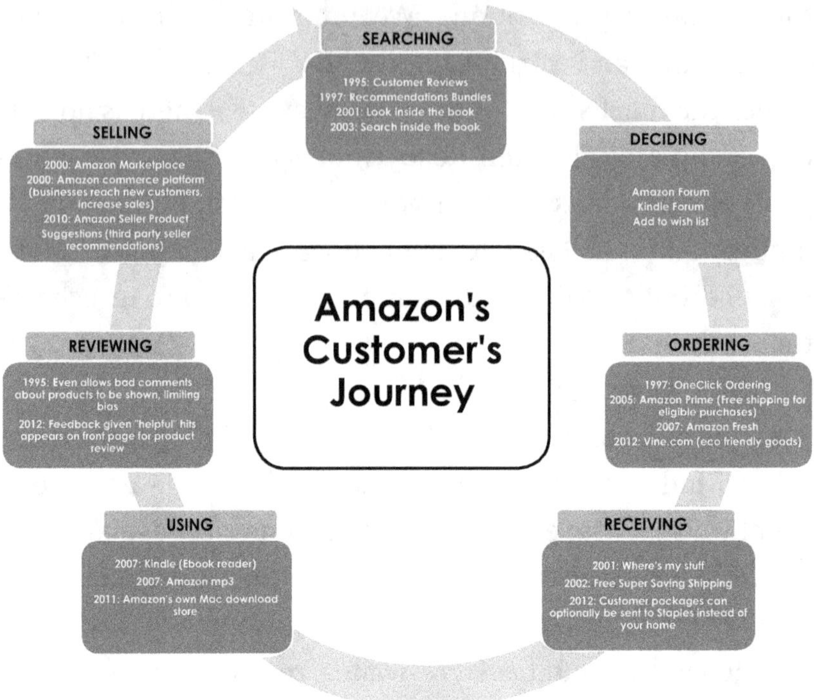

Business Networks: \$t\$ Efficiency. The Fortune magazine article hyperlinked above [4] also said this:

> *One way to serve customers is to give them the lowest possible prices, and in that way Amazon is more like the granddaddy of American retail, Walmart (WMT), than its Silicon Valley counterparts. To offer the lowest prices, Amazon is famously frugal.*

Famous for its network of warehouses around the country (and now strategically situated in overseas locations as well), Amazon runs an extremely tight ship. Bezos is known to have personally fretted about a customer order which was delayed by half a day. Early on, after its inception, Amazon realised the need for back-end capability. It has always distinguished itself from its competitors on its fulfillment capability that gives it an outstanding t Efficiency on key monetary and non-monetary parameters. Starting from a 400 sq. ft. garage storage in 1995, the company had graduated to two fulfillment centers of nearly 300,000 sq. ft. in two years, which grew to a network of nearly 50 fulfillment centers with more than 25 million sq. ft. in 2010. Hiring the best people in the business, Amazon learned to digitise its supply chain rapidly and continues to manage it even more efficiently than its famous competitor, Walmart. A lot more detailed information on Amazon's strategy and its outcomes is available from the following slideshare presentation by faberNovel [5] (http://www.slideshare.net/faberNovel/amazoncom-the-hidden-empire), which I encourage you to pursue to get a complete sense of the key drivers of t Efficiency at Amazon. Let me summarise by saying that the company has managed to effectively enmesh the best of back-office fulfillment capabilities with the best of front-end customer interfaces to create a fast, pleasant and competitively priced offering backed up by a strong infrastructure of owned and contracted capabilities to fulfill the promise.

Amazon is highly data driven

However, innovation and lowest-cost (reputedly Amazon's cost is effectively 20 percent lower than Walmart, on average) are not the end of the story that leads to a growth of $24 billion in three years. In the digitised supply ecosystem that Amazon has orchestrated (I am hoping by now you have pursued the faberNovel slideshare [5] presentation linked above, and have a sense of Amazon's ecosystem), it very early realised that every product line, every product, every customer and every transaction must be profitable in its own right. While investing in the future by selling low cost Kindles, the company is still focused on making sure each transaction is fully analysed for its costs and benefits to the company, and profitability is optimised for each transaction.

How is this possible?

Very early in its life cycle, Amazon acquired and then continued to refine software that enables it to understand the cost for each product. Being a data-driven company, it analyses its cost-to-serve at the minutest level possible to understand which products, fulfillment promises or customers are not profitable. The faberNovel slideshare above gives an example of a best-selling folding chair that took 15 minutes to pack, eroding all its profitability. In 2000, Bezos negotiated with the supplier to send it pre-packaged for an extra $0.25. This strategy of discouraging sales of Can't Realise Any Profits (CRAP) products, combined with its ability to vary the pricing based on customer profile, has now made the company one of the trend setters in TOP.

Innovation at Amazon comes in waves of enhancements

APP involves judicious release of newer versions of products (such as the Kindle) and features on the website (such as one-click purchase) in order to balance the profitability and innovation drive. We will discuss this ability of Amazon in a lot more detail in a later chapter, but Figure 7.3 (page 143) on innovation within the company shows the phasing of Kindle over the past several years.

In November, 2007, Amazon launched the Kindle 1. Although it took about a year to grow, by late 2008, the number of titles in Kindle Store expanded to more than 275,000. But development did not stop there. In February 2009, the Kindle 2 was announced and soon after, the Kindle DX was available. By July of the same year, the Kindle 2 matured and its price dropped. It was reduced once again in October. By mid-October Amazon decided to stop selling the Kindle 2. Even though Kindle products enjoy a short product life cycle, improvements are continually being made and the pipeline of new products is always ready for launch in succession. *(Kindle Facts and Figures).*

Outsourcing and modularisation form key underpinning of Amazon's ability to act

Last, but not least, is Amazon's ability to deploy outsourced services in a results-oriented and modularised manner. The first thing to notice is no single company carries out all the outsourced services. A mix of in-house and outsourced service providers are deployed in a modularised manner to create the outcomes that the company wants. All the service providers, whether its IT services or logistics services or other suppliers, work in a results-focused environment due to t Efficiency drive and TOP. Because the data, know-how and smarts reside with Amazon, the service providers work towards Amazon's strategic objectives rather than their own.

In 2001, the second largest U.S. bookseller, Borders, decided to call it quits online and chose Amazon for outsourcing its operations. Next, Amazon took over the web operations of Borders Online Inc. and re-opened it as a co-branded site. The site was powered by Amazon and on top of that, it handled Borders' inventory, customer service, and shipping services for books, music, and video sales. All and all, this was a win-win situation for Amazon; eliminating a competitor, receiving new consumers and some extra revenue. (Copeland, Lee. "Boarders Turns to Amazon for Outsourcing." Computerworld. 16 Apr. 2001) [6].

By 2002, the company quietly formed an electronic commerce outsourcing business with clients like America Online, Target and Virgin MegaStores. Another significant deal Amazon obtained was with Toys "R" Us in 2000. At this point, the business exceeded more than 30 partnerships. Despite the fact that Amazon wants to develop additional deals, it also wants to continue being selective. Amazon's then vice-president of business development, Owen Van Natta, said that "name-brand companies that are really committed to the customer experience, those are also the partners that tend to really appreciate what Amazon has built." The alliances Amazon formed bestowed on it a large benefit compared to rivals like Yahoo and eBay (Fitzgerald, Michael. "Amazon Finds Profits in Outsourcing." CIO. 15 Oct. 2002) [7].

In a nutshell, Amazon is a powerhouse built on a solid

Figure 7.3: APP in Amazon

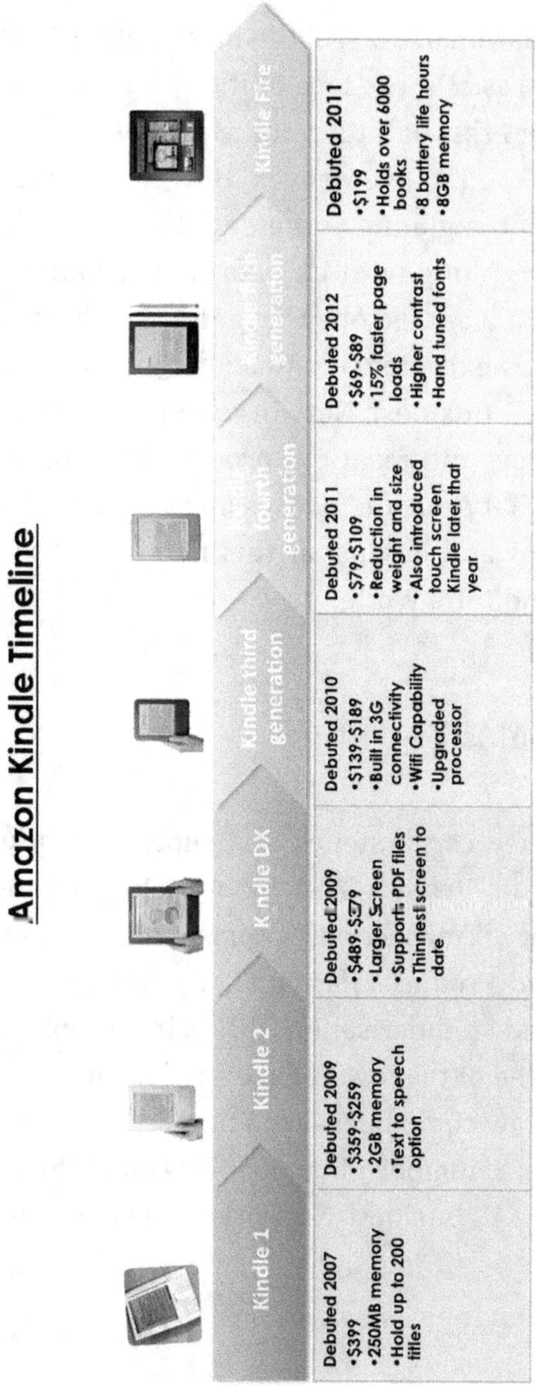

foundation of 5-STAR Business Networks that it crafts and nurtures. On the face of it, the company delights its consumers through its retail websites focusing on selection, price, and convenience. However, what actually makes Amazon a giant are the details we examined in this case study. Along with its enormous product range, Amazon takes advantage of every potential effort to customise the buyer's experience. Amazon demonstrates the power of 5-STAR Business Networks to the fullest. Innovation, efficiency, optimisation, product phasing and outsourcing are what Amazon focuses on every day. While it will have its ups and downs, at this moment, it is clear it will be around for a while.

SUMMARY

Five key cornerstones of a super networked business are Fire-Aim-Ready (FAR) Innovation, $eed-to-$tore (t) Efficiency, Transaction Optimisation Profitability (TOP), Advanced Product Phasing (APP), Results-oriented Outsourcing and Modularisation (ROM). In fact, excellence in any two or three of them will make any business an industry leader. Very few companies do all five of them at world's best level. While a number of companies could illustrate the concept of 5-STAR Business Networks, Amazon excels in all of them.

INNOVATION

Chapter 8

Why Innovation is Not a Solo Sport

> ***It's really hard to design products by focus groups. A lot of times, people don't know what they want until you show it to them.***

Steve Jobs (1955 - 2011), Business Week, May 25 1998

> ***In most people's vocabularies, design means veneer. It's interior decorating. It's the fabric of the curtains of the sofa. But to me, nothing could be further from the meaning of design. Design is the fundamental soul of a human-made creation that ends up expressing itself in successive outer layers of the product or service.***

Steve Jobs (1955 - 2011)

When Apple first launched its iPod music players in November 2001, no one had an inkling what it would lead to in a decade's time. The first salvo that Apple fired in what eventually became a market domination of the mobile phone market, did not look like a mobile phone at all.

Indeed, none of the mobile phone players - whether market leaders such as Nokia or Ericsson or followers, such as Samsung, Siemens, LG, and others - displayed any sign of discomfort at the launch of iPod. The entire battle of iPod and iTunes was fought on the turf of the music industry's drive to save itself from the $0.99 per tune pricing policy of Apple.

Fire-Aim-Ready (FAR) Innovation keeps your competitors guessing

Ironically, an industry which was half decimated by music piracy was busily fighting the very network which would eventually save it - while the industry that would eventually be decimated was cheering on.

When the iPod first came out, all the competing products in the market were clunky, large and erratic in their functioning. Transferring music files from CDs to computer to digital music devices was a major task and piracy and peer-to-peer sharing was rife on the internet. Many observers doubted the wisdom of selling tracks on iTunes when they were freely downloadable on the internet.

With its white headphones, small form factor, easy-to-use interface, user-friendly computer link to both Mac and Windows, iPod quickly became an epitome of 'cool' from 2001-

2006. As successive versions of iPod were issued, (we will cover version update in much more detail in APP), newer features, more capacity, updated iTunes software for computers were introduced, luring the customers to buy many more versions of iPod. Newer products in the same genre - the iPod mini, iPod Nano, iPod Shuffle, iPod Touch - were introduced to further capture the market, each with its own round of successive generations.

Each step in the Fire-Aim-Ready (FAR) Innovation generates multiple platforms for future innovation

Apple was making huge inroads into the hearts and minds of the customers - especially non-Mac users who had no previous experience with the iconic brand. At the same time it was gaining expertise, knowledge and experience in small portable gadgets, learning how consumers interacted with them, learning what were the missing pieces in the existing gadgets and, most importantly, building key connections with the network of suppliers, developers and other network participants who would eventually make the iPhone a resounding success.

Meanwhile, smart phones were becoming popular with consumers and Blackberries were the norm for the corporate markets.

When Apple **fired** its second salvo in the battle for mobile phones, it was **still not fully ready** in the traditional sense. For example, most corporate IT barons scoffed at the idea of using iPhones for corporate emails and networks because of the missing security features. Even the cameras on the first version were far more basic than those on the other smart phones on the market at that time. However, the first iPhone was still an outstanding success. This can be attributes to its massive consumer following from iPods, 'cool' advertising, far superior user experience, reputation for delivering on its promise, and a number of sticky and unique features which were marketed very well.

Behind the scenes, Apple had already assembled an outstanding network of suppliers ranging from Foxxcon - the Chinese assembler to Fingerworks - the developer of the then-unique touch screen software, to the ARM - the licenser of the CPU. Working closely in a huddle with these suppliers, each sworn to secrecy, Apple retained its leadership position in product development time compared to its competitors by outpacing them significantly, while at the same time raising the product quality to levels that leapfroged the competitors. Its innovations were truly disruptive in the industry, even though it always leapfrogged the competitors rather than take the low road to disruption as suggested by Clayton Christensen in his theory. The disruption in the industry is evident from the fact the RIM - the maker of Blackberry - is now on the ropes unsure of its own financial future.

Fire-Aim-Ready (FAR) Innovations build a self-reinforcing ecosystem

Having fired the first salvos in the battle of the mobile phones, Apple moved on to the next steps. It started taking aim towards its final goal of market domination. It expanded its network of suppliers by a factor of hundreds of thousands by releasing the SDK (Software Development Kit) for the Apps for iPhones and iPod touch. Millions of Apps have now been released, each tested and approved by Apple to perform to its standards, making the developers big stakeholders in the success of iPhones.

Innovation is the fire in the boiler of an enterprise

As stated famously by the renowned management guru, Peter Drucker, "Innovation is the specific instrument of entrepreneurship, the act that endows resources with a new capacity to create wealth." An ability to innovate is the **Fire** burning inside the boiler of an enterprise that makes it successful and enduring. So it should not be a surprise that the first cornerstone of 5-STAR Business Networks is related to innovation. You will also discover through reading this chapter why we call it FAR innovation and why the companies that practise this achieve far superior results.

Innovation is a heavily researched topic

To some extent, every business innovates at some time. A new product, a new way of making old products or at least new technologies to build or supply the products or services occasionally show up on the radar screen on most businesses. Many other businesses, such as 3M, have made innovation their hallmark of fame. Most other businesses lie somewhere in the middle of the continuum of possibilities in this respect. The key questions are how do super networked businesses innovate better than the rest of the businesses, and do their results match expectations.

While researching the pace of innovation, I was not surprised to note that no other topic besides leadership has prompted more business books. In fact, after my research associates accumulated more than 550 summaries of business books on the topic of innovation, I had to stop them from collecting any more. It was simply impossible to go through any more of them and be able to discern new thoughts.

Towering above all the other writers and intellectuals on the topic of innovation is the Harvard professor Clayton Christensen. Among his books are the three titles: The Innovator's Dilemma, The Innovator's Solution and The Innovator's Prescription. The titles sum up the content of each of the three books quite succinctly. [1][2][3] All these books are interesting to read and use a number of case examples to set up the dilemma - the eternal struggle of the innovator to stay at the cutting edge of the innovation circle - and then provide the solution - create disruptive innovation by targeting new customers and fend off the competitors. I will not attempt to paraphrase his thoughts on innovation, but rather quote a few selected words relevant to our discussion:

> *"A value network is the context within which the firm establishes a cost structure and operating processes and works with suppliers and channel partners in order to respond profitably to the common needs of a class of customers. Within a value network, each firm's competitive strategy, and particularly its costs struc-*

ture and its choices of markets and customers to serve, determines its perceptions of the economic value of an innovation. These perceptions, in turn, shape the rewards and threats that firms expect to experience through disruptive versus sustaining innovations." The Innovator's Solution - Page 44 [2]

In our experience, the **value network** mentioned above, akin to the 5-STAR Business Network, is almost as important as the firm itself. This is important not only for innovation, but also for efficiency, profitability, product phasing and modularised outsourcing. Importance of the value networks - a network of suppliers and channel partners - to innovation is underpinned by its use throughout this book and, for that reason, I thoroughly recommend you Clayton's book.

Many companies falter in their innovation efforts because of mismatch between the life cycle stage and supply chain structure

Whether the innovation is disruptive or sustaining, whether it responds to the needs of new customers or existing customers, poorer customers or richer customers, or yet a different segment of customers altogether, the process of innovation itself is improved considerably by using the 5-STAR Business Network effectively.

It is now well accepted and established that most products typically go through a product life cycle similar to that in Figure 8.1 on the next page. In the introduction stage, the product concept is put out, research and development are carried out

Figure 8.1: Product Life cycle (Source: Booz Allen and Hamilton, P. Kottler)

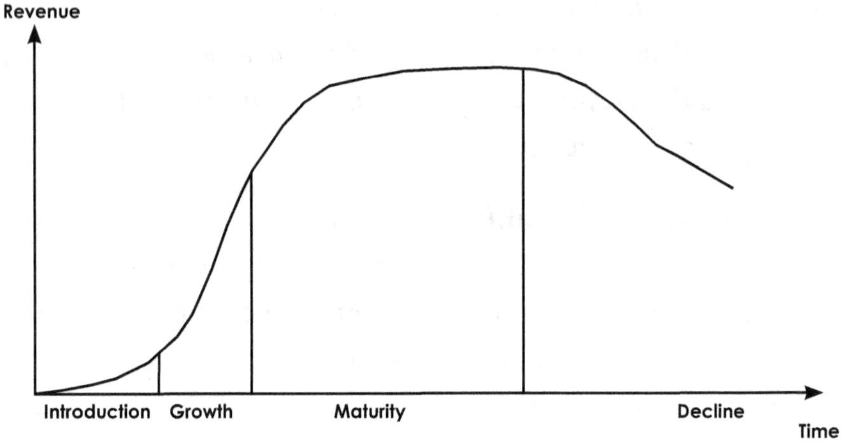

and the product is introduced to the market. At this stage, the key focus of the company is carrying out research and development and introducing the product to the early adopters or to the enthusiasts who have a pressing problem the product solves. However, it is a very critical stage because the early enthusiasts will either like the product and decide to recommend it by word of mouth to peers or they will influence others to reject the product. Also at this stage, the company is still gathering market intelligence about the product features, value in use, and sophistication of the technology with the intention of improving the product.

As the product catches on, it goes into the growth stage, where the early majority starts accepting the product and there is a massive ramp-up in the volumes, in production, and in profits. The company is scrambling to keep up with the demand at this stage. The key focus is to make sure that the shelves are full and stores do not run out of the product.

In the next stage the product hits maturity. The growth starts tapering off and the competition becomes more established in the marketplace. Supply chains are becoming more defined and customers are becoming well-educated in the product benefits, features, differentiating factors and value in use. Now the company wants to make sure that it can supply the products to the customers at the most profitable margin.

Finally, the product enters the last stage of its life cycle called the decline. At this stage the product is finally being phased out either because it has been replaced by another technology or product, or because it is no longer necessary for the lifestyle of the consumers. In this stage the company wants to maximise its own profitability before making a decision to discontinue manufacturing the product.

We have observed that many companies, even multi-line companies, follow similar trajectories, albeit over a much longer period of time. The maturity stage is extended for a long period of time in these life cycles, though the other stages are not inconsequential. Our work in supply chain strategy with growth stage companies has convinced us the supply chain model that is most appropriate for a mature corporation will almost certainly fail in a growth stage corporation.

Supply chain maturity differs at each stage

In fact, our supply chain maturity model shown in Figure 8.2

describes four stages of supply chain, where each stage of product life cycle is paralleled by a stage of supply chain.

As Figure 8.2 below shows, there are four relevant zones of operations determined by two key factors - product maturity and supply chain maturity. Zone 1 is the foundation zone in which both the product and the supply chain are quite immature. As the name implies, in this zone the foundation for the future business is being laid. The next zone on the top left quarter of the matrix – the Innovation zone - implies a relatively mature supply chain, but a developing product. As the name suggests, this is the zone where both product and process innovations are rapidly taking place. The profitability zone on the top right quadrant is where both the product and the supply chain is relatively mature

Figure 8.2: Supply chain maturity model (Source: Baddock and Sood)

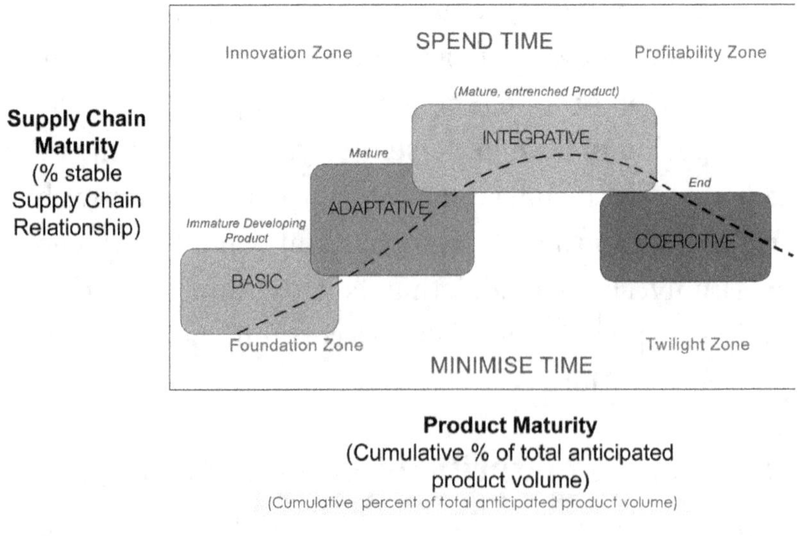

and while incremental innovation might be still possible. This zone is primarily focused on enhanced profitability. Finally, the twilight zone on the bottom right corner is when the product is reaching the end of its profitable life cycle and the supply chain becomes brittle.

Profitability zone and innovation zone - the two most productive zones

Needless to say the more time spent in the profitability zone the more a company can reap rewards of its efforts. However, to maintain fresh product lines, and constantly stay on cutting edge and retain long-term leadership, companies will have to also spend some time in the innovation zone. Intuitively, companies want to spend time in the top two quadrants and minimise their time in the bottom two quadrants. In fact, overlapped on the four zones is a typical supply chain maturity cycle we observe. We will discuss this conundrum in more detail in Chapter 11 where we observe the APP strategies of the 5-STAR Network businesses.

Initially, in the introductory stage of the product life cycle, the supply chain is still very basic. In this improvisational stage of supply chain, the key focus of supply chain team is to really gather enough material somehow, from somewhere, to make the product or to keep the research and development team supplied with raw material. They are not doing any advanced planning at this stage. They are not even aware of all the raw materials or all the parts, which will be required for making this product. A Bill of Materials may not exist or, if it does, it is incomplete. There

Table 8.1: Supply Chain Maturity Model

SUPPLY CHAIN MATURITY MODEL	Basic/Improvise	Adaptive	Integrative	Coercive
Corresponding Technology Life cycle Targets	Innovators	Early Adopters	Early Majority	Late Majority
Corresponding Product Lifecyle Phase	Introduction /R & D	Growth	Maturity	Decline
Key Supply Chain Intent	Gather material somehow for R & D and initial runs	Get the product to customer on time and within budget	Make a delivery promise (based on STP-segmentation, targeting and position of the customers market) and keep it at optimum profitability	Maximise profits in the short term anyhow, as the product is phased out
Focus and Intensity of Innovation	Very high product innovation, no process innovation	Some product innovation, very high process innovation	Low product innovation, some process innovation	No product innovation, low process innovation
Key Supply Chain Control Mechanism	None	Delivery Schedule and Budget	Balanced Scorecard	Profit and loss statement
Supply Chain Systems Sophistication	Little	Some	Very high	Little
SC Agility	Very high	High	Medium	Low
Supply Chain Focus	None	Effectiveness	Effectiveness and Efficiency	Efficiency
SC Personnel Profile	Operational-react to request for purchasing	Leaders-juggle multiple priorities simultaneously and achieve effective results	Managers-create and maintain structured approach, gradual improvement in structure	Executors-short term focus on profits for own company and nothing else
Supply Chain Collaboration (with partners)	None	Forming	Highly evolved	Dissolving
Supply Chain Organisation Structure	No structure	Flat	Matrix	Hierarchical
Planning Time Horizon	Very short	Short	Long	Very short

Figure 8.3: Basic/Improvised supply chain

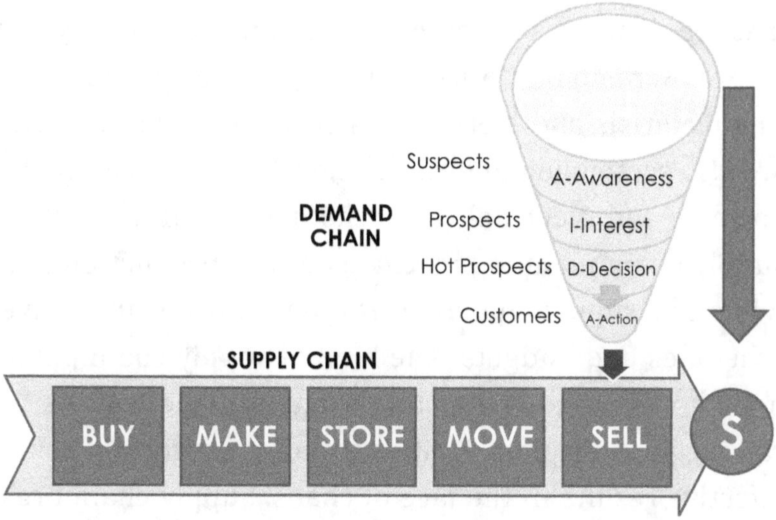

is no supply chain planning mechanism besides this Bill of Material. There is no supply chain control mechanism either. Even a budget does not yet exist, or it might be just very rudimentary. At this stage of supply chain maturity, the companies are not worried about its efficiency at all. There is no supply chain collaboration with its partners for this simple reason: We do not even know who they will be.

Rapid growth in the second stage puts a lot of pressure on the supply chain - and needs an adaptive response

In second stage of product life cycle, as the product moves to the growth stage, the key focus of the supply chain team now changes rapidly. The product has caught on. There is massive demand and ramp-up in production. Supply chain team is scrambling to keep all the factories supplied with raw materials and stores supplied with the product. They are also scrambling to form relationships with supply chain partners and meet delivery schedule within budget. Key supply chain planning mechanisms are production plans, while key supply chain control mechanisms are delivery schedules and budgets. The kind of supply chain personnel who are required at this stage are true leaders who can juggle multiple priorities simultaneously and achieve effective results in the face of chaos. Supply chain organisation structure is relatively flat and the planning time

Figure 8.4 - Adaptive supply chain Structure of EGL

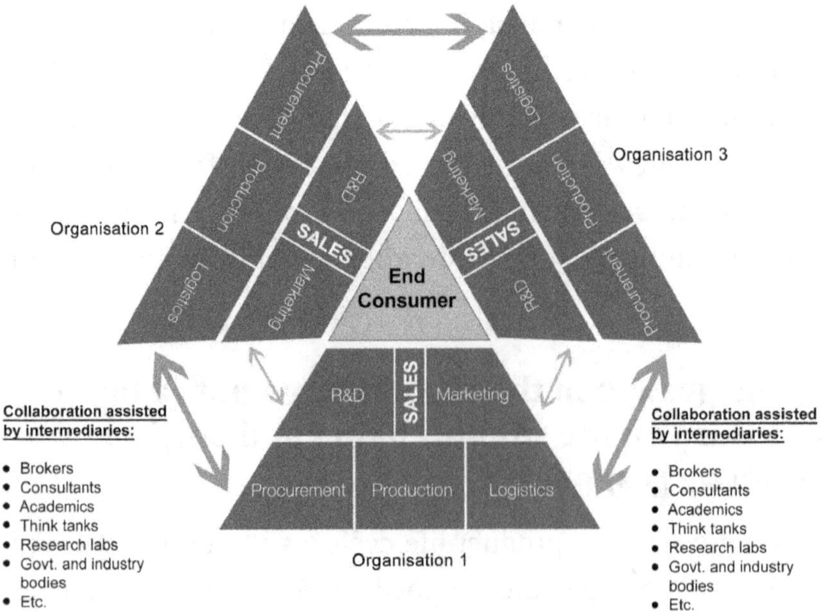

horizon for supply chains is relatively short. You cannot plan for five years when you still need to focus on meeting demand for the next season. As shown in Figure 8.4, adaptive supply chains involve formation of collaboration effort between various supply chain partners to meet the end-customer demand. At this stage, collaboration is crucial - not only in production and procurement, but also in R&D, marketing and other functions. None of the partners has the knowledge, resources or even experience to produce the product alone and it would take an enormous amount of time and investment if one organisation attempted to do it all by itself.

Third stage is all about maximising profitability

In the next stage of the product's life cycle when the product hits maturity, the supply chain team breathes a sigh of relief. They are no longer rushing to meet a very unpredictable, rapidly growing demand. On the contrary, the demand is now volatile. It goes up and down, depending on the month of the year or day of the week. Now the time has come for a collaborative approach to supply chain planning. Collaboration is highly evolved and the focus of key supply chain personnel is now to make a delivery promise and keep it at an optimum profitability.

Key supply chain planning mechanism is an integrated supply chain plan (see Figure 8.5 on the next page), which incorporates highly formalised plans including demand, supply, inventory, transportation, warehousing, produc-

Figure 8.5: Integrated Supply Chain

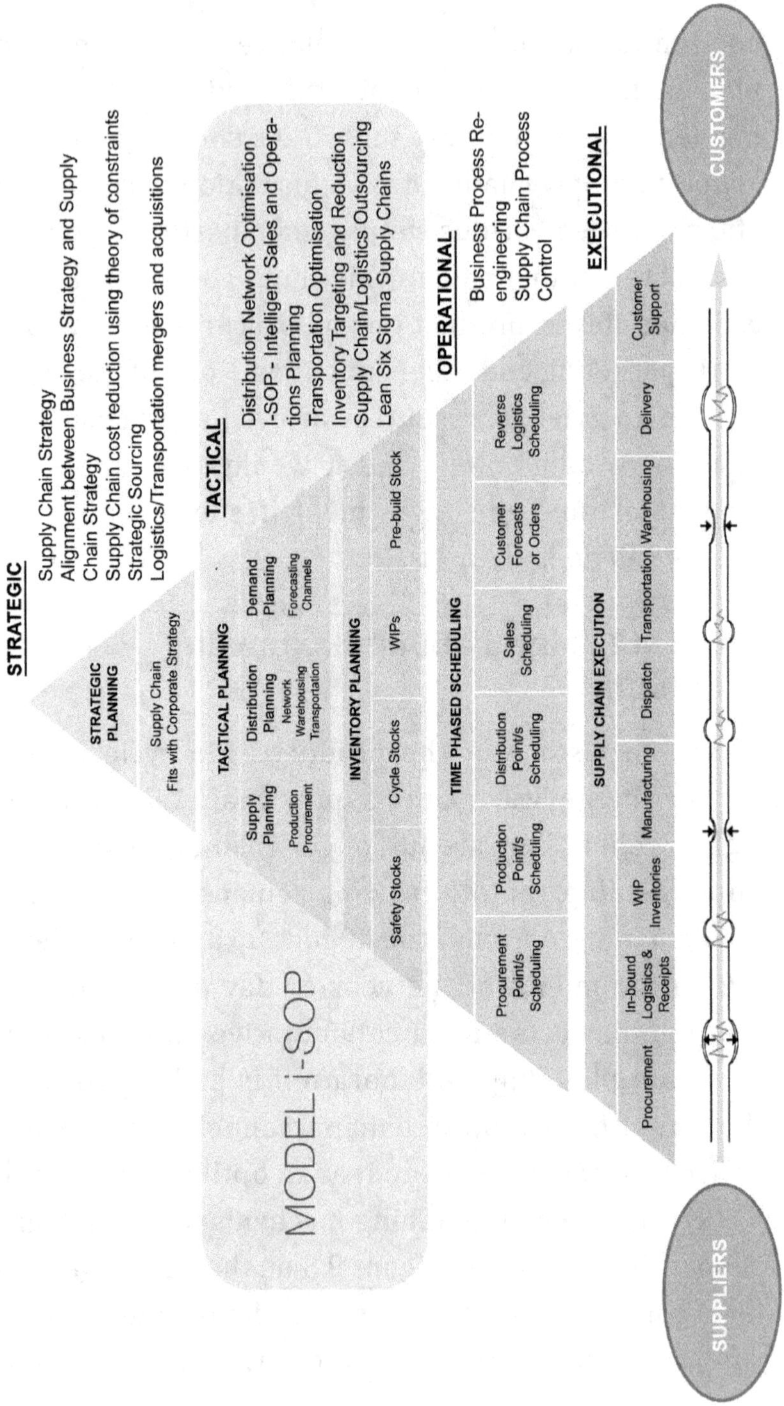

tion and schedules at all levels. Now the supply chain efficiency is at a peak. Agility has gone down but collaboration is high. The key supply chain personnel at this stage will be managers who can create and maintain a structured approach in the supply chain, and who can gradually improve the structure.

The supply chain planning time horizon is now quite long, sometimes running for years and the organisation structure is matrix structure. We will discuss this type of supply chain in detail in the next chapter - t Efficiency.

In the last stage, a slow decline is managed and sometimes reversed

The last and final stage of the product life cycle is the decline stage. The business wants to really phase out the product in a profitable manner and supply chain collaboration partnerships which have been established for years will start dissolving. The focus is to make sure that profitability is not hurt as the product declines in its uptake. Key supply chain focus is to maximise profitability as the product is phased out. The key planning time horizon is now quite short and the type of people who are required are executives with very short-term focus on the bottom-line. The key supply chain planning mechanism is now back to production plan; control mechanism switches from balanced scorecards in the earlier stage to a profit and loss statement now.

Fire-Aim-Ready (FAR) Innovation involves three steps

Now we have seen these four stages of supply chain maturity that correspond with the four stages of the product life cycle. Why is this important?

Firstly, many supply chain managers of high growth organisations are rightly skeptical when told by 'experts' that they should make their supply chains more efficient, formalised and/or integrated. Intuitively they know that in the high growth phase an adaptive supply chain is much more appropriate. External parties with experience in mature stage corporations with integrative supply chains come in and point out lack of equal degree of sophistication without realising the corresponding difference in maturity of supply chain or organisation life cycle.

Secondly, as the supply chains expand into supply networks with 5-STAR capabilities, they can simultaneously foster the efficiency of the integrative supply chains and effectiveness of the adaptive supply chains - as we discussed earlier.

There is no doubt that there are three legs underpinning the stool of capability to innovate - creativity (ideas), customers (marketing) and execution capacity (supply chains). Without these three the innovation dies early on.

A number of authors propose a simplistic three-or four step linear process towards innovation: generate idea, develop product, market it. For example, in "Sustainable Development for Engineers", [4] a somewhat complex text edited by Karel Mulder, the authors show a simplistic six-phase lin-

Figure 8.6: Linear Innovation Model (Source: Mudler et al) [4]

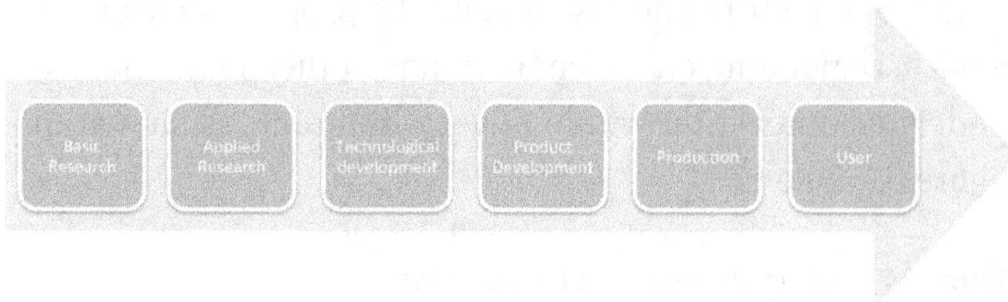

ear model (given in Figure 8.6). I agree with the authors that to some extent this type of model represents the reality, yet they are too simplistic to be of any value.

There is no start of the innovation without creativity. One of the earliest proponents of innovation, Schumpeter (1934), believed that it covered the introduction of a new product or a new quality of a product, the introduction of a new method of production, the opening of a new market, the conquest of a new source of supply, and the carrying out of a new organisation. With this far-reaching definition, the scope of innovation is multiplied immensely. However, a new conundrum arises; what comes after the initial idea? The customer or the execution capability?

This is not idle speculation. One of our large corporate clients faced exactly this dilemma when investing in the renewable energies sector.

Innovation can be driven either by demand or supply

In simple words the dilemma was whether or not an entrepreneur with a good idea should go looking for customers or go looking for the ability to build the product/deliver the service. Depending on the choice made at this point, the two models of innovation will look very different, as shown in Figure 8.7 below:

Figure 8.7 - Alternate models of innovation

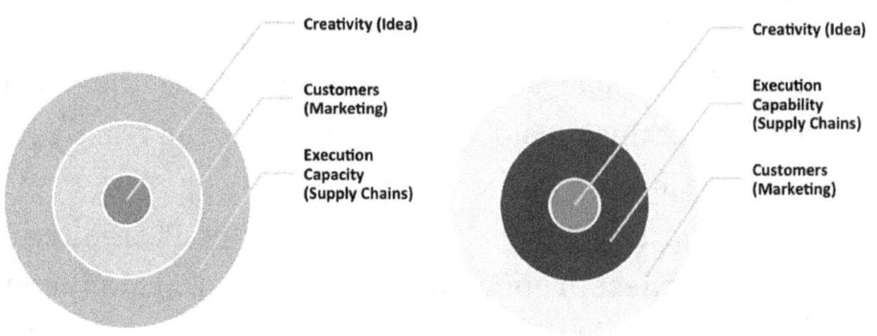

Alternative 1 – Traditional Economics   Alternative 2 – Supply Side Economics

There are enough proponents for both types of models. In our work with clients this conundrum has frequently surfaced and is always argued quite passionately by different senior executives in either of the two camps. On one hand, there are the proponents of the Reaganomics supply-side theory. They argue: 'if you build it, they will come'. On the other hand, the more traditional thinkers argue that, unless we have the customers and services pre-defined, how can we build anything?

Objective analysis of key drivers of demand and supply will show the basis of innovation

At one location, as a way out of this dilemma, we listed the factors on which the decision was dependent. Among the factors were questions related to the certainty of the customer demand, such as:

- Are the real customers properly identified?
- Do the real customers know exactly what they want?
- Have the real customers communicated their demand to the market explicitly?
- How likely are real customers to change their minds?
- How easy is it for real customers to change their preferences?

There were several factors related to the level of innovation itself including whether it was just an incremental innovation or a giant leap. Key questions in this realm were:

- By what factor (multiplier) does this product/service improve the customer life?
- What are the existing means of getting the same value in use?
- What makes the new product better than the existing product?
- What tangible measures can be used to measure the superiority of the new offering?

- Do the real customers actually value the new offering as superior as the provider does?

Fire-Aim-Ready (FAR) Innovation will only succeed if supply chain structure matche demand and supply drivers

To better scrutinise these factors, we developed the following matrix at Figure 8.8 (on the next page):

Red box (box 1) is the danger zone

In box 1, (red colour) in a situation where the level of innovation is only incremental and the certainty of customer demand is very low, massive investments are required for customer education and for building a viable supply network. This is because most potential suppliers will see the situation as high risk and will only respond to monetary inducements to buy co-operation. A very high percentage of innovative efforts are in this square and, as a result, fail because of lack of deep pockets and difficulty in fighting the battle on two fronts simultaneously. A basic supply chain would be the only possibility in this instance where the organisation has to move on two different fronts to build demand and supply simultaneously, as shown in Figure 8.3 on page 159.

The key lesson for the players in the red box is that you need very deep pockets to fight the battle on both fronts - demand and supply. If you lack that financing ability, try

Figure 8.8: Innovation Drivers and Supply Networks

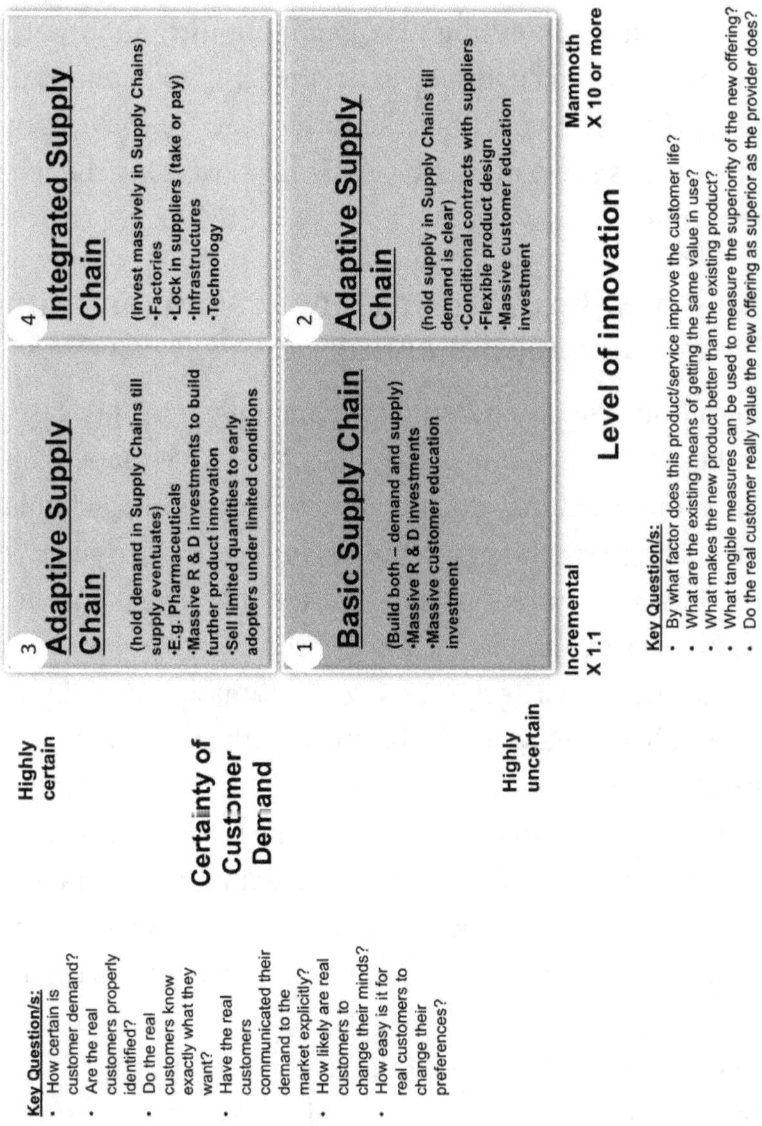

and move up or right - either find a niche of customers with pent-up demand looking for the right product/service by moving up, or move right in the matrix by creating a step change in the users' lives by creating a product that far surpasses anything else available in the marketplace in terms of the customer experience. If you cannot do either of these two things, keep looking for ways to make one of these two moves or consider scrapping the idea altogether.

Blue box (box 2) needs more customer intimacy and/or supplier agility

Recall our client case study; the company in this case was in box 2 (blue colour) - they had a massive innovation but uncertain customer demand due to competing technologies promising a similar magnitude of innovations. In such a case, the business suffers from a chicken and egg situation. The customers do not buy because of some uncertainty, perhaps regarding which technology will ultimately win the battle (after all, no one wants to be stuck with a Betamax VCR and find that VHS standard has won the battle). At the same time, the company is not in a position to invest too much in production capacity unless the customer demand is certain. Most businesses in this type of situation try and work with customers on a conditional basis - promising to build capacity if the orders are placed. However, customers are not inclined to place orders because of the factors driving uncertainty in their own world. A way out of this situation then is to work with the supply network on a conditional basis - promising and delivering massive returns as the demand materialises. The key is to find the right suppliers with world class capability and flexible capacity who are willing and able to understand the situation and work in it. At the same time, flexible product design and investment in customer education to reduce demand uncertainty and increase buy-in also yields good results.

The need for flexibility, adaptability, ability to hold supply in readiness for demand that builds up through education, clarity and events result in an adaptive supply chain that looks like alternative number 2 in the round figure (Figure 8.7) on page 166. The key strategy is to make sure that suppliers and co-developers of technology and production capacity are fully on board with the plan and work alongside your business - as part of your 5-STAR Business Network. If you have any doubts about any of the co-developers or suppliers, it is better to continue looking, negotiating and influencing until all the members of the 5-STAR Business Network are fully on board with you. Consultants, think tanks, industry organisations, academia, research laboratories, and brokers play a critical role in bringing together businesses that could form part of the same 5-STAR Business Network. They play an even more important role in keeping the network humming smoothly, and ironing out any wrinkles in the relationships. Such an adaptive supply chain is shown in Figure 8.4 (page 160).

We have already seen how, in such an adaptive model, several organisations work together in an adaptive network to think and solve problems for their common customer/s. Our client, an entrepreneur with massive innovation, used this model to work alongside some of the largest and best heavy machinery and engineering corporations in the world to bring their technology to the market successfully. This was also a good example of the FAR Innovation, but we will use yet another case example of a much more ubiquitous product, the iPhone, later in this chapter to illustrate this effect.

Orange box (box 3) needs more supplier intimacy and/or customer flexibility

In the box 3 (orange colour) in Figure 8.8, the situation is exactly the reverse. Imagine a pharmaceutical company trying to find the cure for cancer. As the population ages, the demand is already present and growing. However, on the supply side of the equation, the research and development is being carried out in the laboratories of large pharmaceutical companies and their collaborating partners in academia, scientific establishments, consultancies and other organisations. The patent system, to some extent, restricts the collaboration, barring this anomaly every company would be keen to collaborate openly to gain part of the returns of a first mover advantage in a blockbuster product. However, in any scenario, an adaptive supply chain similar to the figure above will result in far quicker and more effective innovation at a much lower cost. In fact that is the reason for collaboration, despite the patent laws. Successful strategies in this scenario will hold demand in supply chains till supply eventuates. At the same time, the business will make massive R&D investments to build further product innovation and sell limited quantities to early adopters under limited conditions - which is indeed the case in the pharmaceutical industry.

We will not discuss box 4 (green box) in this chapter because such a scenario, where demand is highly certain and the level of innovation is mammoth, is rarely encountered.

Fire-Aim-Ready (FAR) Innovation needs an effective business network

We have seen how the super networked businesses use their 5-STAR Business Networks to build adaptive supply chains and gain massive advantage in the field of innovation. But this does not happen only in the fields of pharmaceuticals or high technology. Apple, Amazon, Inditex and many other case studies dispersed throughout this book demonstrate clearly how super networked businesses innovate better than others in their industry.

Consider the case of Apple once more. Whether it is the iPad or Apple TV, the company has never shied from firing first and then taking aim towards the target. For example, the much maligned Apple Newton, a total flop was released around 1995, served as the key platform for eventual success of iPads. Just because the technology or the public was not ready for the product, Apple did not shy away from testing, learning, improving, testing again, learning more and eventually succeeding. As it succeeded with other products and learned the lessons (establishing iTunes and the iPod before launching the iPhone), it had less flops along the way. Eventually Apple will be ready with a unique, highly personalised and anticipated experience for each customer - the Holy Grail of the modern era. Most successful companies have followed similar

FAR trajectories to innovation. Discussions of more examples will make this book entirely about innovation, and there are 4 other cornerstones to our 5-STAR Business Networks that need to be discussed.

I will conclude this chapter by quoting the excellent book by C.K. Prahlad and M.S. Krishnan called *"The New Age of Innovation - Driving Co-created Value through Global Networks"*: ***"To win in the competitive landscape defined by creating one consumer experience at a time, decision makers must develop a whole new mindset for understanding their global supply, logistics and communication networks."*** [5] 5-STAR Business Networks discussed throughout this book enable you to do precisely that.

SUMMARY

Innovation, besides marketing, is traditionally reputed to be the key engine of growth. Product innovation involves newer, better products and services; process innovation involves newer, better way of making or delivering products and services. As companies, products or services innovate, they go through a known life cycle and business network capabilities must be matched to the requirements on the life cycle stage in order to maximise effectiveness. Winning innovators stand the traditional model of innovation on its head by launching products into markets and learning from the customer feedback. The latest example of this is Kindle Serials, which allows authors to write books one chapter at a time while taking customer feedback from previously released chapters. I call this model of innovation FAR innovation. Supply and business networks must correspond to the market realities and innovation prowess of the company.

AA 99914975
Australia
100
AA 99914975

Chapter 9

Cash Rules

> ***There are a lot of things that go into creating success. I don't like to do just the things I like to do. I like to do things that cause the company to succeed. I don't spend a lot of time doing my favorite activities.***

Michael Dell (1965 -)

When Michael Dell started the computer company that bears his name in 1984, the C2C cycle time in the industry was 56 days on average, and as much as three times as long was accepted in the industry. In the book "Direct from Dell: Strategies that Revolutionized an Industry", authors Michael Dell and Catherine Fredman describe the journey of phenomenal growth. Michael Dell started the company convinced that by selling PCs directly to the end-customers, he could better understand customers' needs and provide more customised solutions. Revolutionising an industry was not what Michael Dell set out to do, though it ended up happening once Dell fully understood and applied the power of the business network he created.

Cash is still the lifeblood of the business.

No business has transformed its cash cycle as well as Dell. At the time of writing this book, Dell had lost a lot of its shine due to its ineptitude in FAR Innovation and in ROM. It remains to be seen whether the company can recapture some of its lustre. All the same, Dell is still a seminal case study in building a super network of business based on \$t\$ Efficiency. For this reason we will still use Dell as the instructive example in this chapter.

How powerful was that business network? Figure 9.1 shows the outcomes of how Dell created this efficient business network.

In many businesses the C2C cycle, which is a precursor to our \$t\$ Efficiency concept, is the primary measure of busi-

Figure 9.1: Dell's Business Network: \$eed-to-\$tore (\$t\$) Efficiency

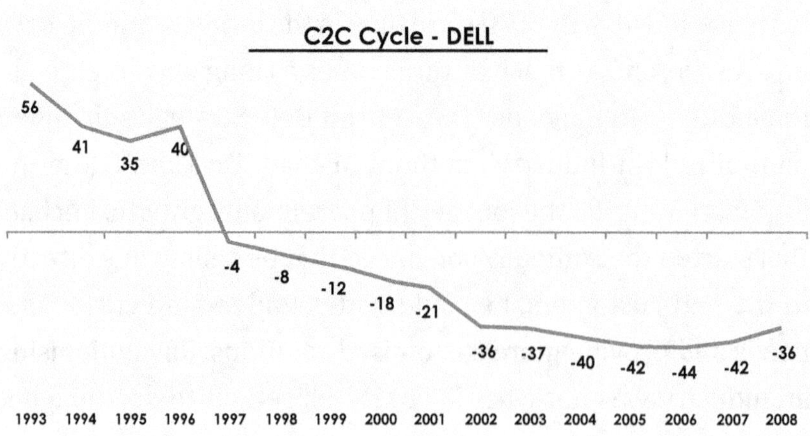

ness efficiency. It measures the average time elapsed between the business paying its suppliers for raw materials and its payment by its end-customers for goods and/or services.

In the PC industry, when Dell started his business, C2C of 50-150 days was the norm. Within four years of starting his business, he had reduced it to -4 days, an unheard of achievement. In fact, Dell was being paid before he bought parts for his product. In the next five years, he brought it down to nearly -40 days, a figure has remained consistent ever since.

If you have not already calculated the bottom-line impact of C2C reduction for your business, divide your annual turnover by 365 days and then multiply that by the cost of capital, obsolescence, and collection. You will be surprised by how high the number is. In Dell's case the number worked out to several hundred of millions of dollars, a figure that easily allowed Dell to consistently take market share away from his high cost, fixed infrastructure competitors.

While in the past five years, the basis of competition in the PC industry has moved to innovation and TOP, Dell still continues to set the standard for t Efficiency.

Discipline and process innovation are the keys to Dell's transformation

How did Dell do this? In his own words [1]:

> *"First, velocity, or the compression of time and distance backward into the supply chain and forward to the customer, will be the ultimate source of com-*

petitive advantage. Use the Internet to lower the cost of developing links between manufacturers and suppliers, and manufacturers and costumers. This will make it possible to get products and services faster to market than ever before.

Second, efficiency and execution will be at least as important as products and services. Turn over to outsiders operations that aren't central to your business advantage; this will free capital to invest in activities with higher value to your customers. How easy you make it for customers to buy and receive your products or services, and the economies you are able to provide to them by effectively harnessing the Internet, are just as important to online brand differentiation as what it is you have to sell.

This leads to the third and most important lesson we have learned: the web experience must be better than any experience in the physical world to create truly sustainable customer relationships and loyalty. Those companies for which the quality of experience has never been part of the customer attraction - companies like discount drugstores and auto parts stores - will be particularly vulnerable to the onslaught of the internet ventures where shopping not only offers a rich and low-cost selection, but also convenience and personalized service as well."

The first and second points above are very pertinent to the topic of this chapter. The third point is more pertinent to improved customer experience through innovation - a topic for the last chapter. The story, as told by Dell in his book, is a fascinating history of how a single man with a mission and idea built a super networked business that still thrives despite numerous copycat moves by the suppliers who have, by now, morphed into formidable competitors. That also explains why Apple continues to pursue court cases against its suppliers such as Samsung, when they allegedly step over the legal line.

Apple has learned from Dell's experience!

$eed-to-$tore (t) Efficiency is the foundation of good business practice

Consider the graphic shown in Figure 9.2 on the next page. We use the example of the retail business networks only because they are most visible. A similar analysis in any other industrial sphere will offer a picture just as revealing.

Consider Amazon, which holds an item on its shelves (in warehouses or stores) for an average of 33 days before a customer will purchase. At the same time, Amazon pays suppliers in 108 days on average, enjoying a whopping cash surplus of 75 days.

Walmart is equally efficient in keeping its inventories churning through the stores. In fact it achieves exactly the same number, *viz.* 33 days, in terms of average days on shelf (in warehouses or stores). However, Walmart is also more efficient in paying its suppliers, within 39 days - enjoying a healthy cash surplus of 6 days on average.

Now let us look at the two businesses most affected by Amazon's startling rise: Barnes & Noble and Borders. On average, both of them pay the suppliers before the customers

Figure 9.2: t Efficiency in retail business networks

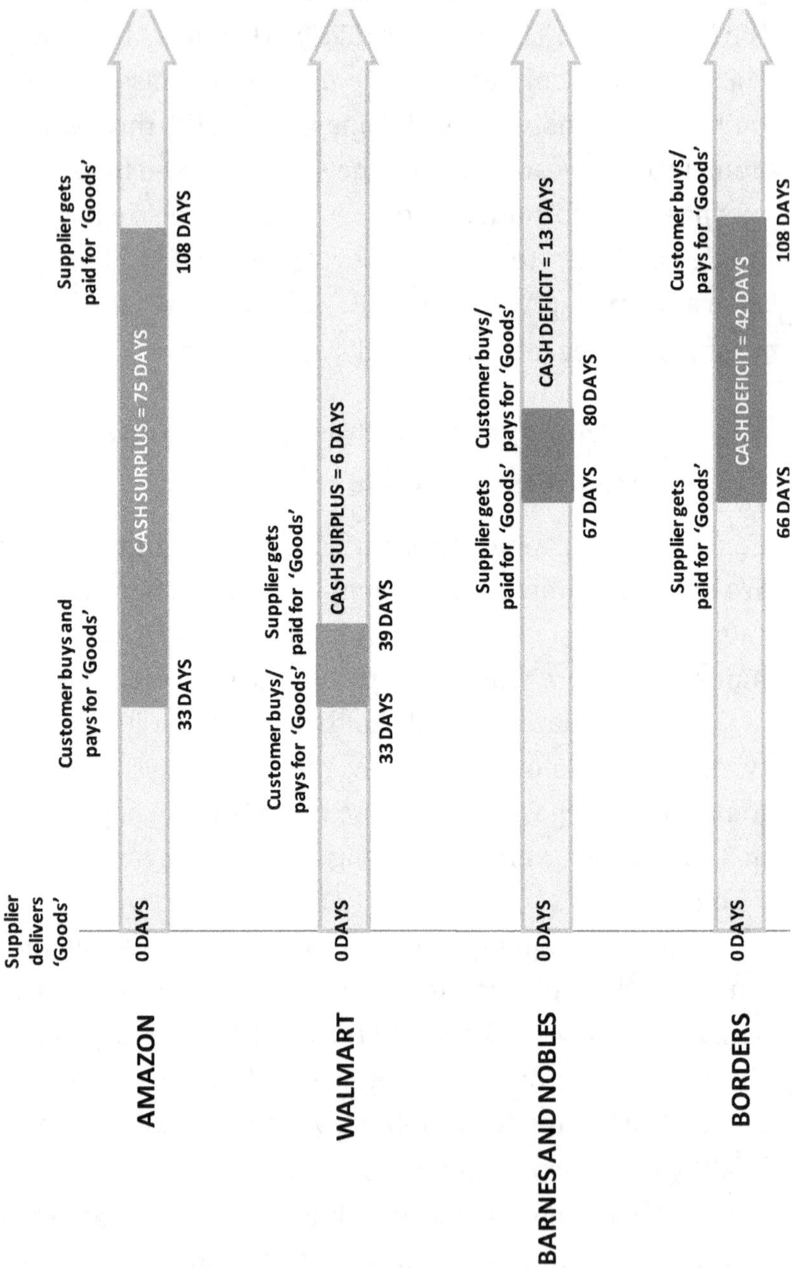

walk to the store and purchase. Because they have not been able to achieve the t Efficiencies of Walmart or Amazon - these companies suffer from a cash deficit of 13 days and 42 days respectively.

How much does the cash surplus or cash deficit actually matter?

Price to Earnings (PE) Ratios tell the story of market expectations

To answer that question, let us look at the PE multiples of each of these four companies.

As is clear from Figure 9.3 below, at the time of writing Amazon is commanding a market price that equals nearly 200 times its annual earnings, while Walmart commands a

Figure 9.3: Outcomes of $eed-to-$tore (t) Efficiency in retail business networks

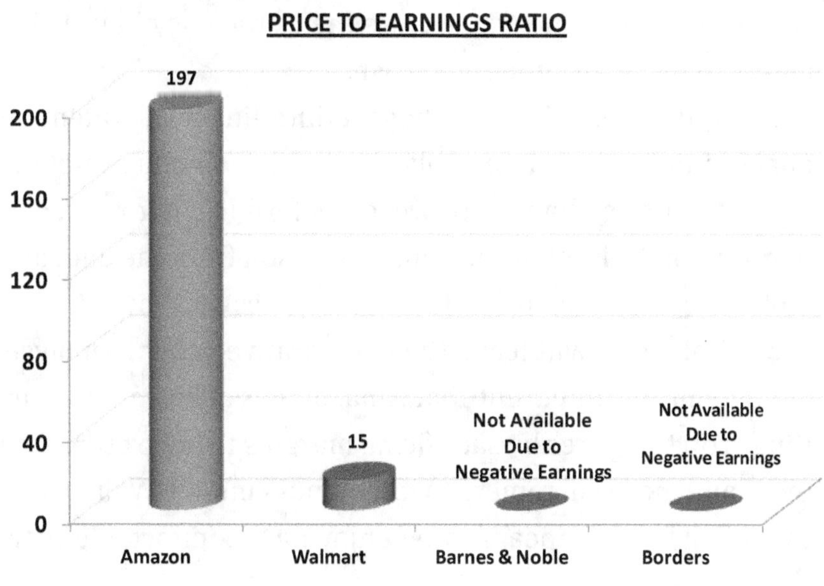

Figure 9.4: Outcomes of $eed-to-$tore (t) Efficiency in retail business networks

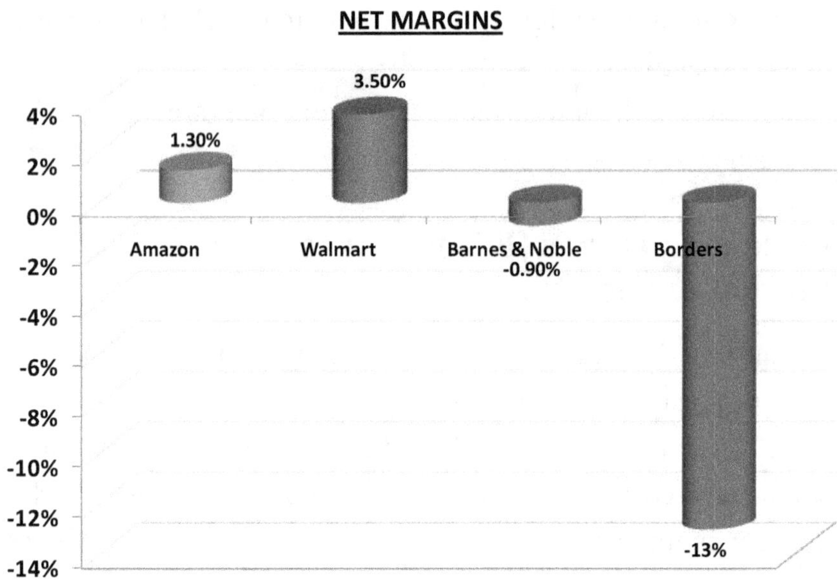

price nearly 15 times its annual earnings. That reflects the market perception of respective growth potential of the two companies, which heavily factors their efficiency in utilising assets to turn cash into more cash.

It is quite possible to get bogged into financial statement analysis and retail sector analysis - which is obviously not the purpose of this chapter. However, before leaving this comparative analysis let us take one last lesson from the data and look at their respective net margin figures as shown in Figure 9.4 above. I will leave the comparative growth analysis to you, but it is apparent that Amazon has achieved its startling growth figures by sacrificing margins to keep attracting more and more customers to its business model. Walmart on the other hand, appears to be enjoying a comfortable margin, a situation which may not last very long if Amazon has its way. It remains to be seen how this battle pans out in the next decade. Barnes & Noble appears to be barely hanging in there, but the future of Borders was sealed by these numbers.

$eed-to-$tore (t) Efficiency goes hand in hand with innovation and both are driven by smart information technology

As we saw in the last chapter, super networked businesses are excellent at innovation. But that is only the first of the feathers in their caps. The other, equally important feather, is the t Efficiency beyond belief. As we saw in Chapter 6 (Figure 6.5 on page 120), when the traditional corporations unbundle into centers-of-excellence based on their core competencies, and then re-bundle into super networked businesses, they acquire the knack of managing the entire network infrastructure in a super efficient manner. A big part of that ability is enabled by information technology and the internet as we saw so clearly in Chapter 5.

Imagine a farmer who buys the seed, tills the soil, plants the seeds, nurtures the crop, applies the fertilisers and pesticides, harvests the crop, sorts and bring the product to the market and finally gets paid for all this systematic effort of co-ordinated activity - doing the right thing at the right time. While the technology and scientific advances have reduced the activity time as well as waiting time in many parts of the process and made it much more efficient, there are still some long lead times in the overall activity chain - no matter how efficient it gets. Super networked businesses continue to challenge both the activity and waiting time; yet they make it as efficient as possible within the existing technological constraints. How do they do that?

In an outsourced global business environment control has given way to influence

It should be noted that only a small portion of the overall activity chain is under the direct control of the business. Let us look at this concept in a systematic manner. Recall the figure below reproduced from Chapter 8 earlier. There is a series of activities involved in the conduct of every business - broadly categorized as per the figure.

Some of these activities are under more control of the business than other activities. For example, the act of buying or

Figure 9.5: Business activities in demand and supply chains

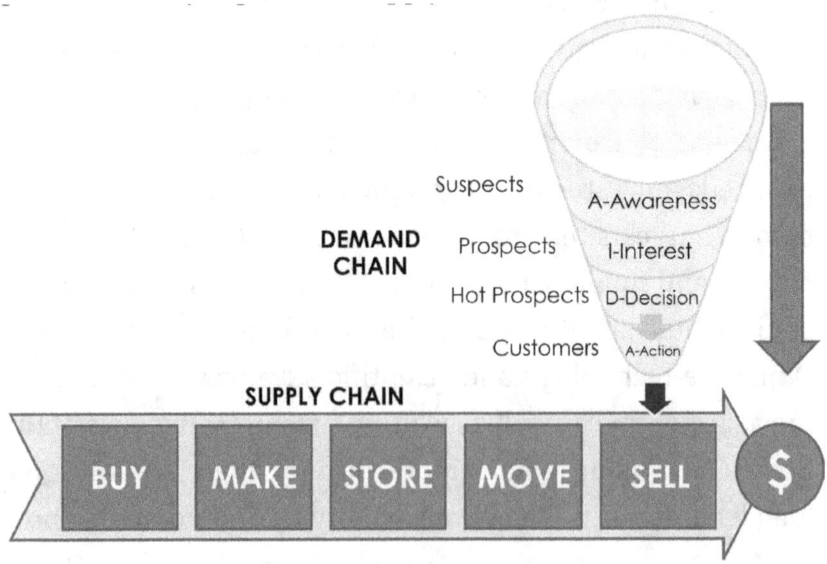

selling by its very definition, is controlled equally by both the parties so the business control over these two activities is only 50 percent each. Other activities, such as production, involve in-house activities, metaphorically within the four walls of the company, so the business control over these activities is nearly 100 percent. Many other activities are either outsourced, or depend too much on factors outside the direct control of the business and hence they range from 20-70 percent in their control factor. This is shown in Figure 9.6 below.

Not all processes get the management attention they deserve

However, as shown in Figure 9.6, the management focus on the activities rarely matches up with the level of control or with the need for efficiency. Traditionally, because manufacturing and sales/advertising are the most visible

Figure 9.6: Business Activities - Focus and Control

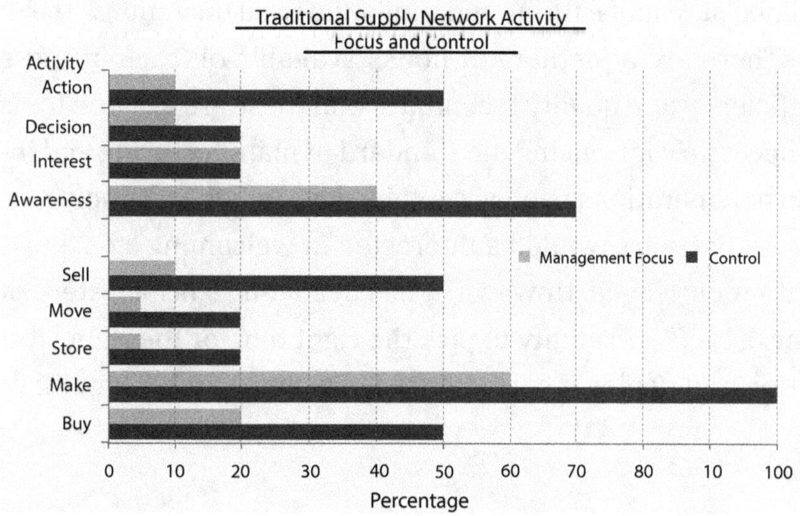

parts of any organisation - these activities garner far more focus and attention than they warrant. On the other hand, store, move and even buy activities get far less attention and focus than they warrant. Part of the reason is over-reliance on the outsourced service provider, due to an assumption that outsourcing activities also outsources the outcomes. We will discuss this aspect of outsourcing later, in Chapter 12.

Most companies focus too much attention on what they can see and not enough on what is invisible

Far more efficiency effort is spent on controlling manufacturing and operations within the four walls of a business than on orchestrating the overall business network. Almost all efficiency techniques, starting from time and motion studies in the 1920s to W. Edward Deming's 14 Principals of Total Quality Management, to Six Sigma and Lean Manufacturing, were primarily aimed at manufacturing and similar internal operations.

There are a plethora of books available on each of these efficiency and quality techniques and these are all good tools - necessary for raising the standard of manufacturing and internal operations to an acceptable level at which supply network partners would be interested in welcoming a business into their league. However, it has been long since mastery of these tools - an ability to pick the right tool for the right situation and deploy it correctly - has become a ticket for entry.

$eed-to-$tore (t) Efficiency takes the game to an entirely new level

Successful network businesses play the game at a much higher level of efficiency. The classical example that illustrates this point is that of the Dell Computers as shown above. To some extent, almost all businesses pursue a similar strategy to garner efficiencies - though not all the companies are as successful as Dell.

A systematic and comprehensive approach is required for the purpose

What is the strategic framework that underlies this t Efficiency drive that has become so common? And why do many companies still fail in their attempt to achieve it? Those are two remaining questions that we will answer in the rest of this chapter.

Figure 9.7 on the next page shows the strategic framework used in countless projects. t Efficiency is consistently one of the key aspects of nearly every project we work on.

The pipeline at the bottom of the pyramid represents the flow of material from suppliers through the company to their customers. A number of stocking point locations are shown all along the pipeline, each with its own variability of demand. A couple of these stocking points show bottlenecks in the pipeline, which are quite common occurrences in most material flows.

The lowest layer of the pyramid - the execution layer - shows all the activity right from purchasing through inbound logistics and manufacturing to storage and customer service.

Figure 9.7: $eed-to-$tore (t) Efficiency - Strategic Framework

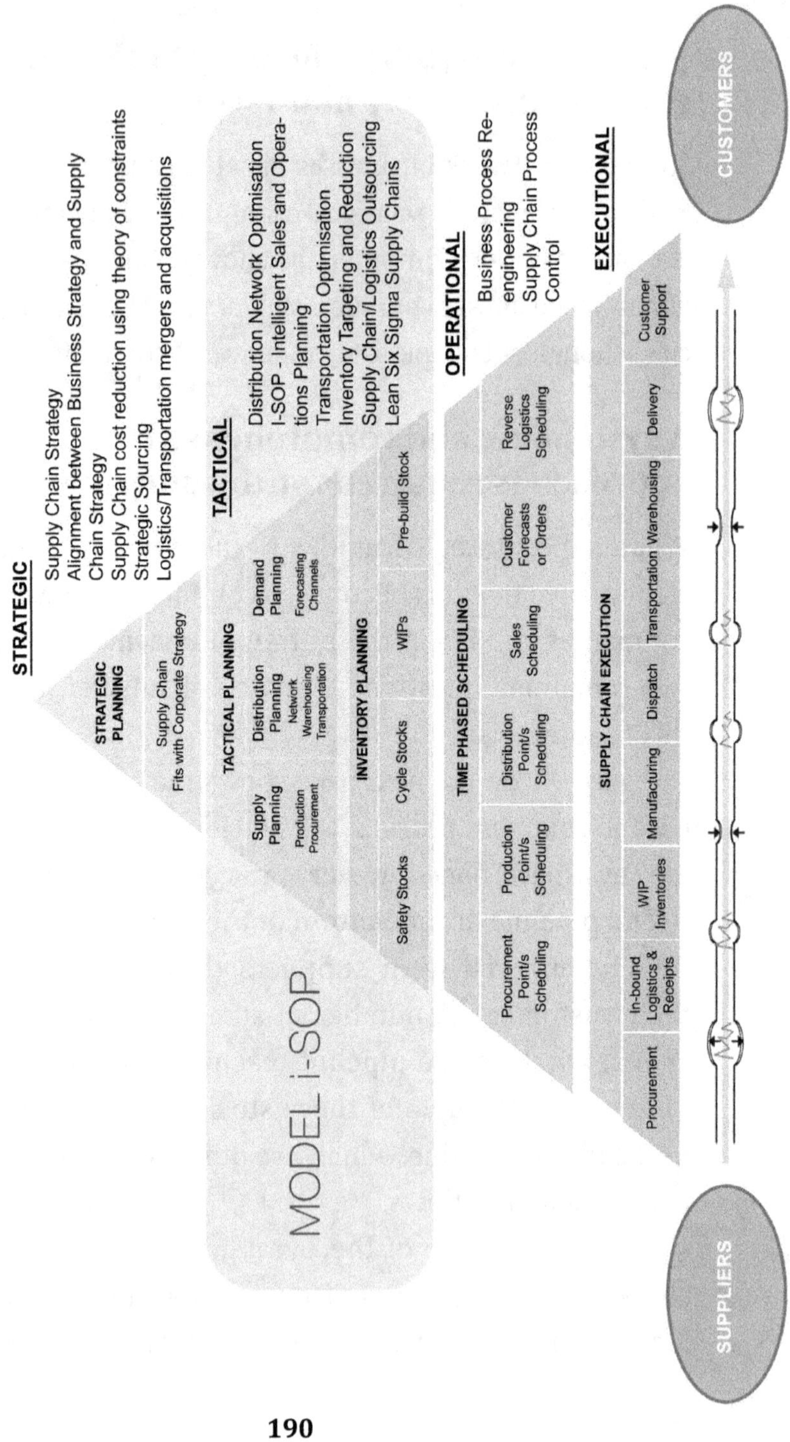

These activities are carried out in a well-orchestrated succession to make sure material flows in, through and out of the business as quickly as possible, increasing the velocity of the supply chain. Moreover, a business can work with its supply network partners to compress the time, allowing it to turn over its cash faster and increase overall profitability of the business. In other words, time compression for the activities leads to a lower C2C cycle and higher profitability, which is, after all, the goal of all this efficiency drive.

In order to move the material faster through the pipeline, most companies spend some time scheduling the activities so that they happen in smooth succession. The time horizon for scheduling does not go beyond a few days (at most a couple of weeks) depending on the level of granularity and detail involved in the scheduling process. Obviously, this time phased scheduling needs to happen at a number of activity points all along the pipeline. While a number of scheduling software systems are available to assist the schedulers, a certain amount of fine-tuning is almost always required in order to come up with realistic schedules.

Each layer builds on the foundation of underlying layers and enhances the efficiency of the whole system

At this operational planning layer, the key means of achieving efficiency are re-engineering the business process and the supply chain process control mechanisms to measure schedules against the actual execution results. While the systems are vital to carry out these functions, it is even more vital to ensure that these systems are measuring the right activity in the right manner and reporting the outcomes against schedules in the right way.

Most of the problems that show up in scheduling and execution are directly attributable to lack of tight tactical plans that link the demand planning with supply planning.

Inventory and spare production capacity are the buffers that allow almost uncontrollable variability in the demand to match somewhat controllable variability of supply. The area shaded in grey in Figure 9.7 on page 190 is the sweet spot of the t Efficiency effort. Dell, as well as any other company that excels in this area, devotes extraordinary amount of time, effort and resources to these activities - and minimises wastage as a result. A full description of all the activities that form the part of Tactical Planning layer is beyond the scope of this chapter. In any case, each of these activities, whether demand forecasting, or production planning, or inventory targeting are far too technical to discuss in detail in this book.

Demand for every product and service fluctuates depending on the customer mood, weather, advertising and whole raft of other factors that drive demand. Every business endeavours to make sure that a ready and willing customer does not go empty-handed. That would not only lose the sale and consequent profit, but also cause a customer backlash against the company from disappointed consumers. Not only that, a customer who temporarily switches over to a rival due to unavailability of your product might start liking the rival's product and switch loyalties. A continual dance of satiety ensues between the demand and supply where it is profitable to produce just enough and not too much. Intelligent Sales and Operations Planning is all about making sure that the two sides of the equation can match each other at most points of time with minimal expediting or waste. In other words an optimisation exercise is carried out to maximise the demand fulfillment and minimise the cost at the same time. We will see in the next chapter how Super networked businesses are going far beyond traditional Advanced Planning and Scheduling towards Transaction Optimisation Profitability where each transaction - a sale, service event, or even steps along the process - is measured for its profitability to optimise the overall business income.

The strategic layer is the most important one

At the top of the pyramid, long-range strategic plans for the business are matched with the long-range operational and supply chain plans. High level constraints in the system and structure are identified and strategies are formulated to overcome these constraints. Various options are discussed and assessed for their risks, costs, and benefits. This sets the stage for the long-term t Efficiency drive. All this may sound like a lot of work but when it is done well, the results create a competitive advantage that lasts for years, if not decades.

It is quite possible to get bogged down in the finer details of JIT, TQM, SCOR and many such toolkits that can help in achieving efficiencies. We know and use many of these tools when required for the specific purposes within businesses. But I will not cover them in this chapter because this is a 'big picture' book. Those who need to know about these tools will already know about them and know which books to buy to learn how to use them. Most do not need to know more about these tools besides knowing they exist and can be useful in specific situations.

We will close this chapter with another quotation from Dell [1]:

> *When we work with our suppliers, we act as an advocate for the customer who uses our products day in and day out, and experiences success and failure in his own business as a result of the things that we do or don't accomplish with our suppliers. It is incumbent upon us to see that our partners respond to market demand so that we'll all succeed - and with a freer flow of information going from the customer, through us, to the supplier, there's a much greater chance of that happening. (page 179)*
>
> *...In our business, virtual integration - the idea of integrating distinct businesses so that our partners are treated as if they're inside our company - evolved naturally out of our need to glean even-better information from our customers and enhance logistics management with our suppliers. ...in marrying the benefits of a tightly coordinated supply chain with the specialization afforded by our virtual connections, we have created what some are calling "a new model for the information age." (page 185)*

SUMMARY

t Efficiency pertains to the management of cash flow through the business network that allows the entire business network to thrive sustainably. Dell is regarded highly in management of its C2C cycle, a vital component of the t Efficiency. Dell revolutionised the PC industry by lowering the C2C cycle from the industry norm of 56 days to a negative number. Most super networked businesses utilise their 5-STAR Business Networks to stand the industry dogma on its head. In the retail industry, Amazon has achieved far more efficiency than its competitors, allowing it to offer lower prices and attract new customers achieving extraordinary growth of 100 percent or $24 billion in three years. A comprehensive framework is required in order to achieve t Efficiency. In addition, there are numerous tools such as JIT, TQM, SCOR, etc. that can be deployed on a selective basis.

Chapter 10

Co-Multiply Your Profits

- Use the Full Power of the Global Business Networks

> ***Know how to listen, and you will profit even from those who talk badly.***
>
> *-Plutarch*

When I was training to be a merchant marine captain, I learned how to navigate a ship - essentially an enormous hulk of hundreds of thousands of metric tonnes of steel, through all kinds of oceans and channels. One of the biggest hazards was the current. This unmeasurable entity could make a very large ship slowly but steadily drift well off-course. If left unchecked you could literally end up on the rocks. In fact, that is partly what happened to the tanker Exxon Valdez and the Italian cruise ship Costa Concordia; currents are a contributing factor in a great majority of maritime disasters.

We learned to counteract the current by measuring the ship's position at two different times and estimate the current using a calculation based on the direction and speed of the ship. After calculating the current in the previous period, it was counteracted in the next time period. The only problem was that over time the current's direction or speed might change. Before the invention of the GPS the only way of finding the ship's position in the open ocean was to measure the altitude of the sun and stars, and this calculation took nearly half an hour to complete. This meant that the intervening time period between two positional observations was quite large and the vessel could drift many miles off course in this period. Moreover, one was always applying a correction that was out-of-date and based on two previous observations. This state of affairs always leads to a ship zig-zagging its way towards its destination. By itself, this was not a huge problem, except in places where real-time information about the sea currents was vital - for example, in narrow shallow waters where minor drifts could leave the ship sitting in danger. Imagine the consequence of tonnes of fuel oil leaking around pristine beaches and sea-shores.

Most business continue to make decisions based on outdated ("out-minuted!") information

I was amused to see that, post-MBA, when I switched to a shore career as a management consultant with Booz Allen and Hamilton, most businesses were still running their management decision-making based on information derived from data pertaining to events that happened long ago. One of the first projects I was involved in during my summer internship exposed me to a company that took nearly four weeks after the end of each quarter to compile its quarterly profit and loss results. Only then were they in a position to make any adjustments in the costs and revenue drivers to meet their budget in the next quarter. By the time the decisions were made and implemented, more than half of the next quarter was already over.

When the margins are comfortable, in a stable industry, not exposed to too much global competition, this state of affairs can work. But today, not many businesses of that nature exist.

IT systems have evolved

Corporate IT systems have made enormous strides in the past 25 years. In my first few projects in the mid-to-late 1990s we were still getting data in dot-matrix printouts from mainframe systems via structured queries. It was extremely difficult to feed and manipulate this data and check it for accuracy and completeness.

The Y2K scare and Enterprise Resource Management (ERP) systems revolution led to massive investments - running into hundreds of millions of dollars for many of our clients - in upgrades of the corporate IT systems. Almost all of the business cases for these massive investments were predicated on the usefulness of the data in decision making. The key question is: after massive investments in IT in several waves do we now have the real time information to make decisions on costs, pricing and profits?

Real-time information is the key to decision making

Before we answer this question, let us first look at what is meant by real-time information and what does it enable businesses to do?

As shown in Figure 10.1, decision-making technology and capability have improved significantly over the past 40 years. Roughly moving in steps of single decades, the capability has expanded in terms of detail of information available and the timing of the availability of the information.

For example, in box 1 of the figure below (the red box), traditional cost and works accounting undertaken over very high aggregate information took a long time to compute profitability of the projects, products or services. The information was recorded either manually or in old mainframe computers with very limited querying capability.

Figure 10.1: Profit information - timing and aggregation

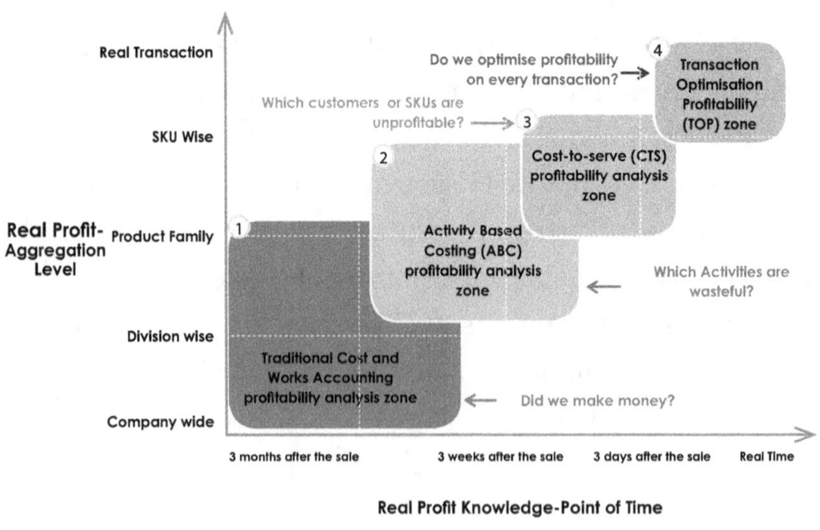

Information manipulation and analytical capability was restricted to human brains or to highly limited computer power. As a result, most information was out-of-date by the time it was available. The key decision makers needed to be extremely well experienced and possess good intuition, and still needed a lot of luck to make the right call, even for routine decisions.

This situation was not very dissimilar to early seafarers navigating wooden ships without any modern electronic navigation aids such as radar, GPS, radio beacons or myriad other equipment. Each new piece of equipment introduced since then enhanced the decision-making capability of the navigation officers and led to bigger, faster and more effective ships as a result.

Now better information is available faster to decision makers

A similar progression of information capability within organisations can be observed over the period of past 40 years. Full exploration of this topic will take up an entire book, let alone a chapter in a fast paced business book such as this. Let us use a table to summarise the findings.

Table 10.1 lays out the progression of decision-making technology and capability in four steps from box 1 to box 4.

Activity Based Costing (ABC) was a major advance in the 1980s

When Activity Based Costing (ABC) became popular in the late 1980s and early 1990s, partly as a result of the Balanced Scorecard Drive, it was seen as a major step forward from traditional

Table 10.1 - Progression of decision-making technology and capability

	Traditional Cost Analysis Zone 1	Activity Based Cost (ABC) Analysis Zone 2	Cost-to-Serve (CTS) Analysis Zone 3	Transaction Optimisation Profitability (TOP) Zone 4
When does the profit information become available to key decision makers?	Between 1 to 3 months after the transaction period ends	Between 2 weeks to 6 weeks after the transaction period ends	Between 1 days and 2 weeks after the transaction ends	Just before the transaction happens so that the decision maker can decide yes/ no
How aggregated is the most accurate profit information when it becomes available?	Aggregated to product family level, or division level and sometimes even company level	Aggregated to SKU level or product family level – based on activity profile	Agregated to customer level, or SKU level – based on SKU/ customer profile	Transaction level – for each transaction
What is the key question asked?	How much money did we make in the last accounting cycle?	Which activities eroded profitability?	Which customers and SKUs are not profitable enough?	How do we optimise profitability on every transaction?
Business Intelligence Level built into IT systems	None	Some – activities wise analytical capability	Medium – SKU and customer wise analytical ability and time-based forecast ability	High – transaction wise analytical ability an time and trigger based predictive ability with appreciation of sources of potential errors
IT Systems	None or Mainframe computers with limited query ability	Client server systems with MRP II type of application for transaction data storage	Distributed Network systems with ERP or enhanced ERP systems for transaction storage and decision support	Cloud servers with best-of-breed applications for data storage and decision support
Outcomes	No real time ability to improve profitability – a bit like navigating in the open ocean without any electronic equipment	Very limited real time ability to improve profitability – a bit like navigating from lighthouse to lighthouse	Some real time ability to improve profitability – a bit like navigating with electronic instruments but no GPS	Continuous real time ability to monitor profitability and optimise it – similar to using GPS for navigation in narrow and shallow waters
Who is the key decision maker?	CFOs and CEOs responsible for aggregated decision maker; a well trained sea captain who can use sextants and chronometers to 'fix' the ship's position	Cost accountants and the process re-engineers - responsible for activities rationalisation; Narrow focused navigating officers trained in terrestrial navigation	Marketing strategists and analysts - responsible for SKU rationalisation; Generalists with good knowledge of modern equipment	Front line employee – salesman or purchasing officer – responsible for the transaction; minimal trained navigator who can read the co-ordinates off the GPS and plot it on the map
Role of the Apex decision makers	Day-to-day decisions because they need experienced persons in the absence of fact based data	Monitor day-to-day decisions because the decision makers lack full perspective on all the facts	Guidance and exceptions based involvement in day-to-day decisions and some direction setting and leadership	Direction setting and leadership – because day to day decisions are made and monitored at lower levels

cost accounting. For the first time, Value Engineering became possible by designing products and services based on only those activities that the customers considered valuable. This revolution was not just a result of the concept of ABC, but also of increased computing power available through the newer client server computers and Manufacturer Resource Planning (MRP and MRP II) applications. In retrospect, there were still many limitations of the decision-making capability with which most are quite familiar.

Cost-to-Serve analytics (by customer or product segments) advanced the ability to differentiate service offerings

As the computing power exploded further, and the mega-systems like Enterprise Resource Planning (ERP and ERPII) were put into service, data capture and storage capability multiplied enormously. At the same time, data manipulation capability grew significantly and newer decision support systems (DSS, with sophisticated data manipulation capability) became available allowing the users to forecast, plan, prioritise and monitor activity. We have discussed these systems in great detail in Chapter 5, where we outlined the role of technology in networks. In theory, these systems should allow the businesses to know the cost-of-service for each customer or SKU instantly by pressing just a few buttons and generating either a standard report or an ad-hoc report. However, in most instances, we have found many problems. An enormous amount of data cleansing and manipulation is necessary before we can calculate the cost-of-service for a customer or an SKU. In many cases, even where the ERP systems have been in use for more than two decades, this is simply impossible without going back to manually linking each transaction to its source paper document and tracing key data associated with that transaction not recorded in the ERP system, simply because the system was not set up to record that data. As the business networks expand and are reconfigured on an almost continual basis, this limitation of ERP type systems is exacerbated. That is the reason why we still continue to see businesses struggling to understand the true profitability of their customers or products, and continue to over-service some customers or carry unprofitable products/services in their offerings.

However, many businesses possess this ability to scientifically differentiate a good customer from a bad customer and to quantifiably differentiate a good product from a bad product in terms of profitability. Either they set up their ERP systems to provide that information with a reasonable accuracy or they continue to tinker with the system or the information until the information is of reasonable quality. While it must be obvious what to do with this information, good companies are very disciplined about how to graduate their products and customers to higher levels of profitability through efficient use of service policies.

The old 80/20 rule applies to customers and products in most cases

How do they do that? For those who have not done this before, let us look at an example of a company that found that while its star 20 percent customers and products contributed to 80 percent of its profitability, there was a sizable proportion of customers and products that were simply unprofitable. When shown as a graph of cumulative profitability these customers and products formed the right most (segment D) value destroying part of the spectrum, as shown in Figure 10.2.

A's are clearly the profitable stars, B's are the solid middle guard, C's are barely profitable stragglers while D's are the clear value destroyers. Many people continue to carry D's only because of their connection with the rest of the business or because sales people continue to make such connections. Many others are determined to get rid of D's at any cost.

All customers and products do not deserve the same service

The company in question, however, embarked on a different tack. It was determined to reward its best customers with the best service. At the same time, it also determined that all products and customers must pay their full way in terms of cost. This meant differential service levels for different customers and products - where vA customers got the best service for A products, and slightly less costly service for B

Figure 10.2: Customer and Product profitability analysis

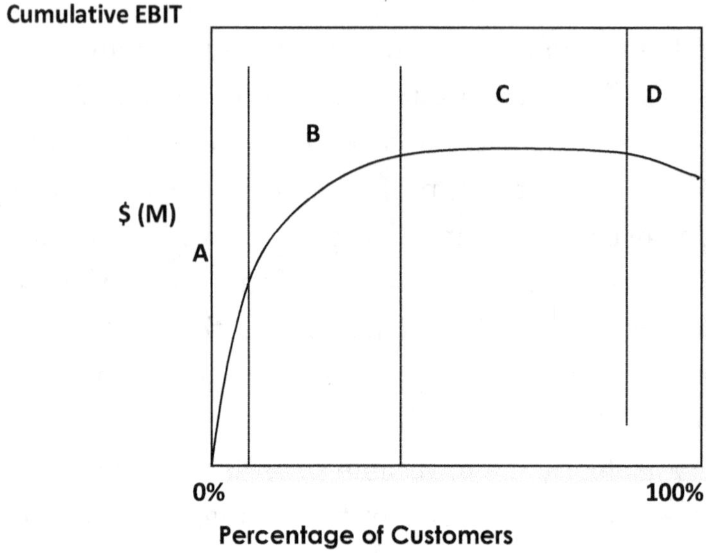

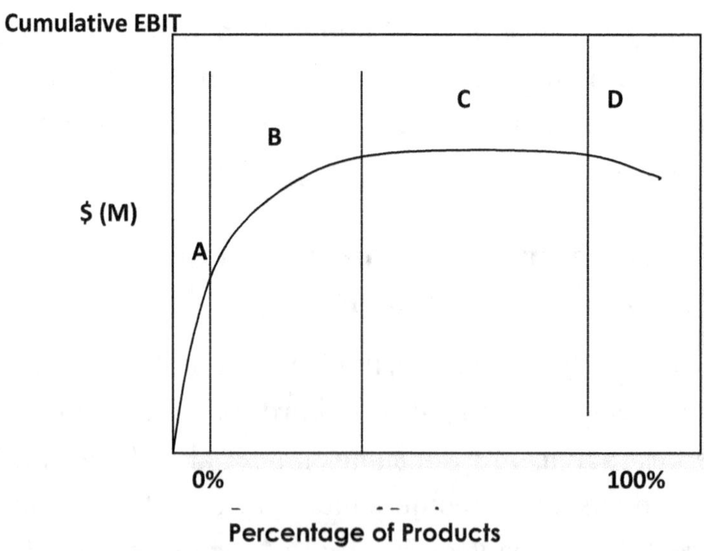

products, and so on. The resulting 4x4 service matrix looked like the Figure 10.3.

Figure 10.3 - Service Matrix - Customer and Product profitability analysis

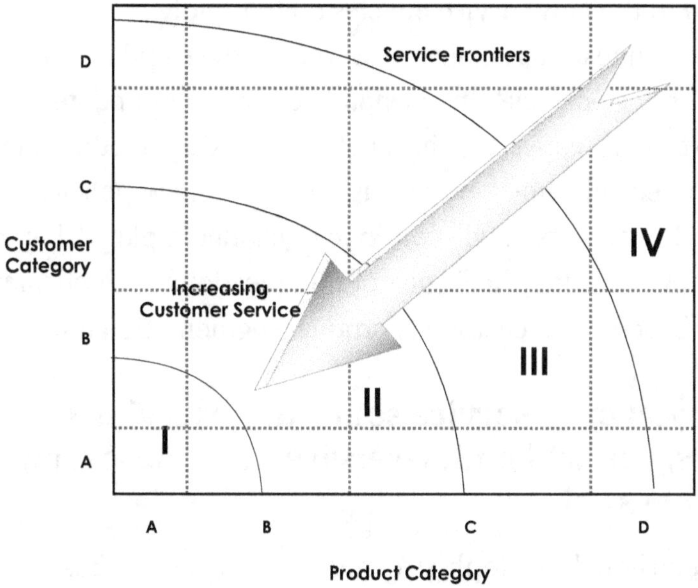

The service matrix can get sophisticated and complex depending upon how many layers the cake contains. The Figure 10.3 merely illustrates four service frontiers based on a relatively simple and static analysis. Service frontier I, for example, might involve same-day delivery in metropolitan areas and next-day delivery in non-metropolitan areas, while the service frontier IV might involve a promise to order the product from suppliers when the customer places an order and communication of delivery time within 72 hours (enough time to order, plan and calculate the lead time).

Obviously, there can be many more service frontiers and the exact definition of each service level will depend entirely on the nature of the business and requirements of its customers. In this case (a specialty chemicals company), the customers ranged from large auto makers to small school labs and most adjusted to the new service policies quite rapidly.

A crucial question remains: how frequently to update the customer service policies? On one hand, too frequent updates confuse customers, especially the smaller ones. On the other hand, as customer behaviour evolves as a result of new policies, the profitability data itself will continue to change rapidly. There is no easy answer, and this is one of the questions each business decides based on its customer/product segmentation mix.

When you over-service some undeserving customers, you will under-service more deserving customers

Consider, then, how sophisticated your own business' cost-of-service capability is and how fact-based your segmentation and service policies are. We use some very sophisticated and dynamic tools for diagnostic purposes, which are particularly useful in organisations where considerable differences of opinions exist on these questions. Many executives and CEOs are surprised by the answers because of the tendency of the subordinates towards selective information dissemination.

Now going back to Figure 10.1 (page 200), while many businesses are not even at level 3 yet, others are moving towards or are already at level 4 - optimising profitability of each transaction as it happens. This happens in many different ways on a multitude of fronts. On the revenue front, customer intimacy has evolved to a level where goods and services are offered to a customer when they most need/

want them at a price that suits them the most. Consider the following example quoted in a recent press report.

Imagine if the customer segment comprised a single customer

Kashmir Hill, writing in Forbes magazine online in February 2012, cited *The New York Times* in an instance of 'How Target Figured Out A Teen Girl Was Pregnant Before Her Father Did.' [1]

> *Target assigns every customer a Guest ID number, tied to their credit card, name, or email address that becomes a bucket that stores a history of everything they've bought and any demographic information Target has collected from them or bought from other sources. [They] ran test after test, analyzing the data, and before long some useful patterns emerged. ... Take a fictional Target shopper named Jenny Ward, who is 23, lives in Atlanta and in March bought cocoa-butter lotion, a purse large enough to double as a diaper bag, zinc and magnesium supplements and a bright blue rug. There's, say, an 87 percent chance that she's pregnant and that her delivery date is sometime in late August.*

The anecdote in the article by Kashmir Hill quoted an instance where a father storms angrily into a target store, demanding an apology for encouraging his teenage daughter to get pregnant by mailing her coupons of baby stuff, only to retract the demand later on when he discovers that she was indeed already pregnant. It demonstrated the power of predictive business intelligence. This article and the New York Times article [2] by Charles Duhigg, and the book it is based on "***The Power of Habit: Why We Do What We Do in Life and Business***" also by Duhigg [3], are worth perusing. They lay out in great detail how companies as diverse as airlines, retailers and insurers routinely collect and analyse an enormous amount of data about the buying and usage habits of their customers.

Sophisticated mathematical modeling of this data creates business intelligence and predictive capability enabling the businesses to make the right offer at the right time in the right manner and at the right price. It also enables the front-line employees to engage with the customers in a transaction discussion or negotiation that allows them to make a sale at a price that maximises profitability.

The companies that are best placed to exercise this type of customer intelligence gathering and predictive offering are those selling their ware online. Online habits are a lot easier to track and trace. Customers find it less intrusive than someone following them around their daily routine. Data is easier to gather, collate, analyse and manipulate. By far Amazon is one of the best companies engaged in customised offerings and pricing.

With real-time information, profitability of every transaction can be optimised in real time

Now let us move on to the cost part of the story of TOP. From a business network perspective, this is an even more sophisticated and esoteric story.

While many companies find it difficult to gather, refine and manipulate data for the accurate cost-to-serve analysis necessary in step 3, others have gone far beyond to registering increasingly sophisticated data attached to every transaction within its business network, with a view to enabling a transaction cost analysis in real time for every transaction. Not only are all the parts of a particular product transparently known, but also all the activities, timing and cost of these attached to the associated service are tracked to enable real-time costing of the product/service bundle so that the salesperson can clearly understand which transactions are not worth getting into. Frontline staff making decisions that drive costs can equally see the product cost impact of their decisions in real time. Not many companies have arrived at this level of sophistication, however, they are clearly driving towards it as they continue to refine their data gathering and manipulating ability within their business networks.

5-STAR Business Network is critical for real time information

So why is the business network critical in TOP? Firstly, the type of data requirements are onerous and can be shared only with selected network partners. Secondly, your network partners need to be equally as sophisticated as your business in their capability to gather, store and manipulate information required for this analysis. Finally, as you work with your network partners to create outstanding products and services through FAR Innovation as outlined in Chapter 8, the network is already primed with information that allows for TOP. Freight logistics forwarders - some of the most sophisticated TOP practitioners on earth today - are almost entirely made up of sophisticated business networks with hardly any other tangible assets. There are several large global freight forwarding logistics companies that thrive because of their ability to maximise their revenues due to their customer intimacy and minimise their costs at the same due to their business network t Efficiency. Their ability to do this for every transaction at the time of quoting for business allows them to maximise their profitability.

In closing this chapter, lest you lose sight of the end goal, we would like you to remember the following quote attributed to Mahatma Gandhi. If you follow his advice, TOP is almost a certainty in today's technology driven business network:

> *"A customer is the most important visitor on our premises.*
> *He is not dependent on us. We are dependent on him.*
> *He is not an interruption in our work - he is the purpose of it.*
> *We are not doing him a favour by serving him. He is doing us a favour by giving us the opportunity to serve him."*
>
> \- Mahatma Gandhi

SUMMARY

Imagine a clever village shopkeeper who always knows how much a customer would be willing to pay for each item they want to purchase, while, at the same time, knowing how much each item cost him, including all the overheads. He will always maximise profitability for each transaction because he makes all the cost decisions in order to minimise costs, while selling products at the highest price possible at all times. TOP moves towards achieving similar results utilising the power of information technology and business networks for large corporations. Activity Based Costing and Balanced Scorecards have paved the path towards TOP. Cost-to-Serve modelling is a cruder form of TOP. More refinement of cost-to-serve models combined with yield maximisation tools will ultimately lead to TOP.

Chapter 11

Do Not Be a One Hit Wonder

> ***"Learning and innovation go hand in hand. The arrogance of success is to think that what you did yesterday will be sufficient for tomorrow."***
>
> *William Pollard (1911-1989)*

When Amazon launched its first generation Kindle in 2007 for $399, the features of the tablet were quite underwhelming. It had a very simple grayscale screen reminiscent of old Apple laptops, a weird looking keypad, 250 kb of memory, no camera, no touchpad, no touchscreen, very limited connectivity via Amazon Whispernet and a capacity of just 250 books. Why did it still sell out in less than five and a half hours? Why did Amazon launch it when there were far more feature-rich products already in the market-place? Chapter 8 (FAR Innovation through 5-STAR Business Networks) answers these questions more thoroughly.

In this chapter, we look at the story subsequent to its launch. The second generation Kindle, aptly named Kindle 2, was slimmer, more powerful and featured 2 GB of internal memory. The battery, as well as the keyboard, were enhanced, and the price was reduced.

Success comes only when the innovation cycle continues in perpetual motion

Since then, Amazon has launched newer generations of products every few months in quick succession, while expanding the product family at the same time. The current product family (correct at the time of writing in June 2012) consists of Kindle, Kindle Touch, Kindle Touch 3G, Kindle Keyboard 3G and Kindle Fire. The feature-rich Kindle Fire retails for just US $199, and the network of services and apps available on this new platform keeps growing. These include more than 100,000 movies and TV shows, thousands of popular apps and games, Amazon Silk (a cloud-accelerated browser), free cloud storage, 19 million songs, email, document reader and a Kindle owners' lending library.

Amazon is not alone in its phased approach to product launch and upgrading techniques. Every major electronics manufacturer, mobile phone manufacturer, and consumer and industrial company follows a similar strategy. Companies with a 5-STAR Business Network are far more successful and profitable in their APP strategy, as we will see by the end of this chapter.

As a company evolves and starts going up the track of innovation, very rarely is the product taken straight through a single life cycle all the way up to decline. Generally, as it goes up to the maturity stage, the company starts thinking about the next generation of the product, and a newer and better product is developed. An innovation cycle begins as shown in Figure 11.1:

Figure 11.1: Innovation Cycle

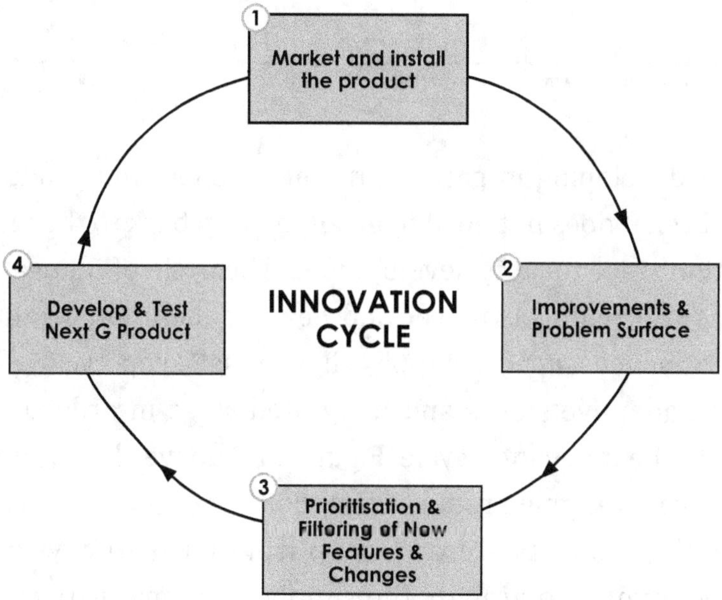

Effort is focused on the current product and future products

While the previous product is still in the growth stage, the company starts developing the next generation product simultaneously. The intent is to launch this next generation product just as the previous generation product starts entering its declining phase. In order to do so, the company

Figure 11.2: Multi-Generational Product Life cycle

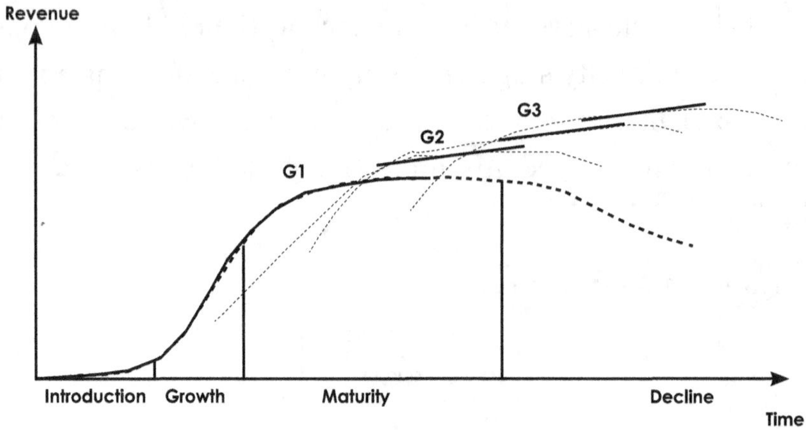

starts developing in parallel, the new generation products. However, it does not need to go all the way back to the introduction stage for this development. The cycle of innovation keeps shortening as newer and newer generation products are launched and the supply chain must keep cycling between adaptive stages and integrated stage in order to facilitate the innovation cycle. Figure 11.2 above shows such a multi-generational product life cycle.

As the products are cycled up from G1 maturity to G2 introduction and growth, the supply chain maturity cycles back from integrative stage to a more adaptive stage.

Supply chains cycle between adaptive and integrative as companies switch from one generation to next

In fact, this cycling of supply chains between adaptive and integrative stages, as shown in Figure 11.3 on the next page, is a key feature of highly innovative corporations that suc-

Figure 11.3: Supply chain maturity model – Supporting innovation cycle

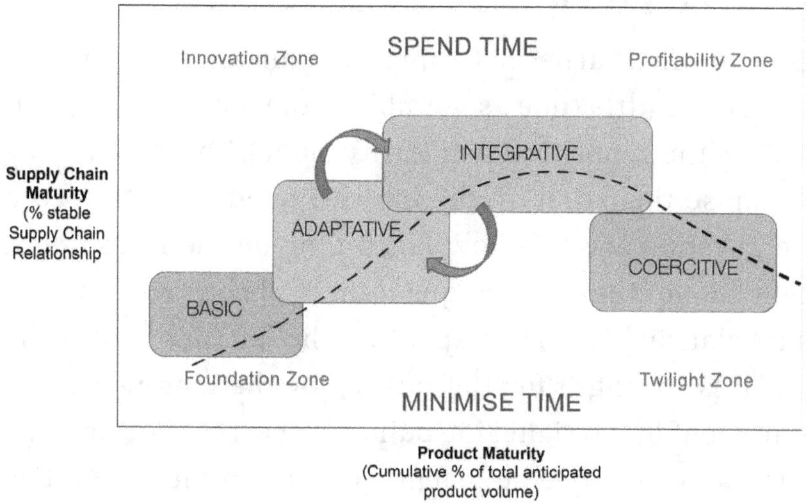

cessfully launch newer and improved products on a regular basis.

It must be pointed out that the frequency of the innovation cycle - how often a company launches new and improved products - needs to be very carefully managed, to balance the demand for fresh innovative products and the short-term profitability derived from the existing product range.

In the growth stage of product life cycle (and during the adaptive stage of the supply chain), the supply chains are not at their most efficient. They are still adapting to all the growth and development. It is in the integrative stage of supply chains, in the mature stage of products and organisations, where supply chains are at their most efficient.

It is a balancing act - between the current needs for profits and future needs for products

As mentioned earlier in Chapter 8, companies would want to spend as little time as possible in the bottom two quadrants of the supply chain maturity model. This means they maximise their time in the top two quadrants. However, the question arises – how much time on the left side vs. right side of the top two quadrants? We believe that this is a balancing act. Time spent on the left side builds the business' strength for the future, for the long-term. The time spent in the right side builds the current bottom line.

If companies give sufficient run time in the integrative stage, they will be able to maximise its profitability, while if it spends sufficient time in the adaptive stage they will be able to maximise innovation. Only by cycling between adaptive and integrative stage in order to maximise both profitability and innovation can a company truly achieve a long-term competitive advantage. In fact, the frequency and speed of the innovation cycle is a finely tuned balancing act, as shown in Figure 11.4:

Figure 11.4 - Balancing innovation and short-term profits

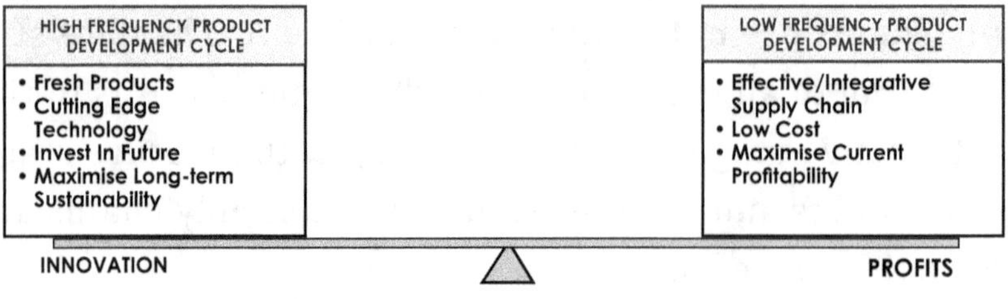

It is unprofitable to launch products too frequently

If the company launches the product too frequently, naturally it will have a very fresh product line-up all the time. However, its short-term profitability will suffer because the supply chain will never have an opportunity to get its maximum efficiency in its integrative stage. Too frequent product launches result in high cost, ineffective supply chain and reduced profitability, as shown in Figure 11.5:

Figure 11.5: Too frequent innovation reduces short-term profitability

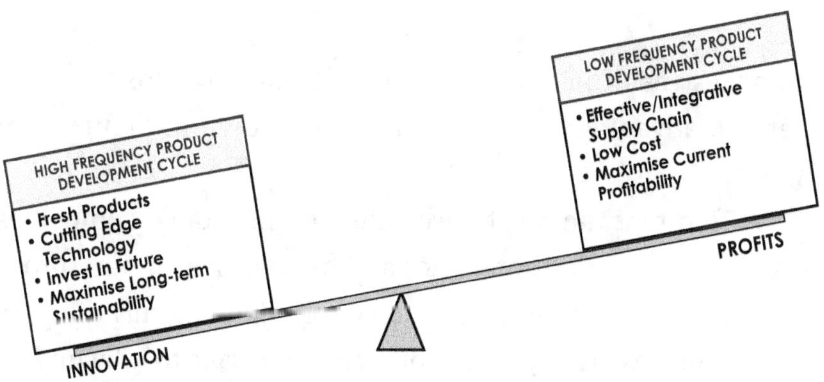

It is suicidal to milk products beyond their useful life

On the other hand, if the company does not launch new products frequently enough, the supply chains will become very efficient and the short-term profitability will be high. However, the companies run the risk that too infrequent product

Figure 11.6 - Lack of innovation reduces long-term sustainability

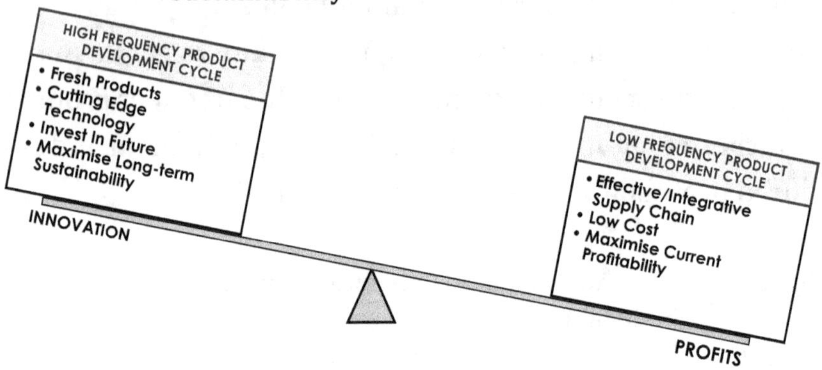

launches result in a stale product, leading to a loss of competitive advantage and innovation as shown in Figure 11.6 above.

We find that the most successful and profitable companies manage their innovation cycles and frequencies. In turn, product launch, innovation cycle and supply chain maturity follow suit. In contrast, many other companies struggle to fully understand the implication of these for their businesses, and how to take full advantage of their supply chains to support their business strategies.

5-STAR Business Networks help take product phasing to a new level

While many companies are very good at product phasing as described above, the super networked businesses do it exceptionally well. They are so good at it in fact that they are a class apart. This is why it is called APP. So what is the magic of 5-STAR Business Networks in APP? Let us see and example to examine how they do it.

One of our clients, a consumer products company, made many, run of the mill, staple products sold every day throughout all the major retail outlets in the country. The drive to efficiency was grinding. Every cent saved over a carton could add up to millions of dollars over the course of the year. In addition, their customers - supermarkets and the retail megastores - continued to demand lower and lower prices every year. All our work with the company focused on driving efficiency benefits through the supply chains using classic integrated supply chain framework. Having shaved millions of dollars in the overall cost base, the company was faced

Figure 11.7: 5-STAR Network allows businesses to be efficient and effective at the same time

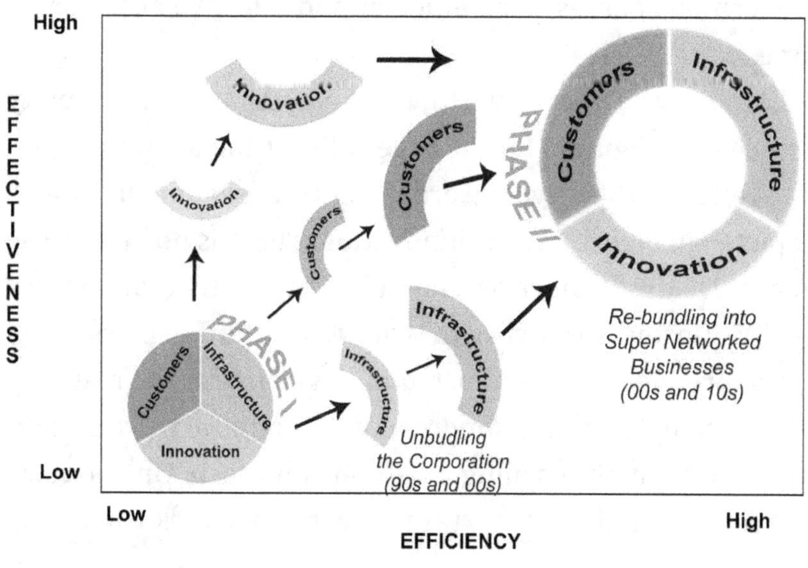

with a conundrum when it started developing a new set of products for high end chains and boutique outlets. The products would not fit in the traditional supply network of the company, so it had to develop a most expensive yet Adaptive supply chain akin to that in Figure 8.4 on page 160 for this product range to meet the customer need.

Internal benchmarking would time and again reveal a vast gap in the cost performance of the two supply networks - yet the overall profitability of both the product ranges was quite comparable in percentage terms. The company was not blinded by the false drive towards cost reduction for the new range and managed to be effective in its product phasing and efficient in its supply chain outcomes.

Recall the Figure 6.5 on page 120 which is re-produced in this chapter (Figure 11.7 on page 223) for convenience of discussion.

We found how a 5-STAR Business Network allows the modern businesses to be efficient and effective at the same time.

Efficiency is the drive towards short-term profitability depicted in Figure 11.4 on page 220. At the same time, effectiveness is the drive towards long-term sustainability also depicted in Figure 11.4 as innovation. There is no better way to achieve this balance as shown in Figure 11.4 than to use a 5-STAR Business Network to work with your business.

On products (or product families) already in maturity stage, the network runs more akin to an infrastructure network - a tight ship sailing on the beam reach. On products or product families in Next G range, the network runs as an innovation network - a sailing ship running downwind on the spinnaker.

Inditex - the company behind Zara - does it well

Perhaps the best example we see of APP is a company known as Inditex. In a highly volatile fashion industry with a stock standard two fashion seasons a year, many of Inditex companies manage five, six or even more fashion season each year. How do they manage to produce so many fashion cycles every year and yet manage to keep the prices low enough for the man on the street to afford it? Amancio Ortega, the owner, truly believes creative imagination is the root of innovation. One quote he lives by, from Walter Lippman, is "When all think alike, then no one is thinking."

Starting off, Ortega realised that it is not only the rich that appreciate expensive and well-designed clothes. As a result, the concept of creating high quality, trendy clothes affordably became Ortega's target. This combination of effectiveness and efficiency has allowed Ortega to become the third richest man on the earth as of early 2013. In fact, today there is no better and more inspiring example of a true 5-STAR Business Network than Ortega's company Inditex. For this reason we will examine the example of Inditex in a great deal of detail in Section 3.

Intel has been doing APP very successfully for a very long time

To demonstrate APP using a business network of suppliers and

Figure 11.8: Results of APP at Intel

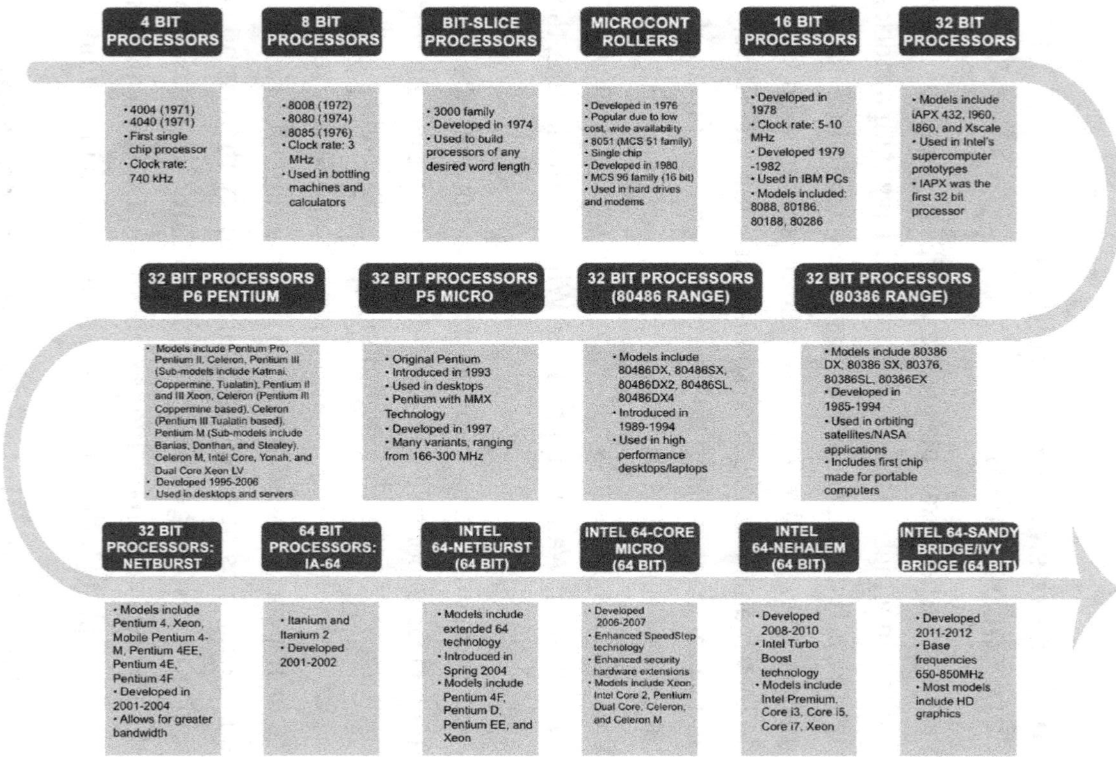

collaborators, we have chosen the example of Intel. Consider Figure 11.8, which shows waves of Intel microprocessors released from the 1970s to 2012.

Intel has managed to do the unthinkable, bringing products to market via product-oriented development and market-oriented approaches. The former focuses on developing the production process or the product itself, while the latter focuses on what customers need. One example of Intel's research and development is the 45nm silicon transistor in 2007, a paramount part of the 'cutting edge' Intel Core 2 processor. This new technology has allowed Intel to move into the cell phone market and the robotics industry. Intel managed to attain somewhat of a paradox within the business: invest in new technology while also continuing to invest in harnessing previous products to maintain a profitable 'product pipeline.' This year, Intel plans to spend $13 billion, of which $2 billion will go toward a microchips development plant, a long-term investment. 'The other $11 billion goes toward more immediate improvements in Intel's manufacturing technology' (timesofindia.com). [1]

One of the reasons Intel is so aggressive on capital spending is to maximise the chances it has of protecting its gross margins as it moves into smaller and lower priced CPUs. Overall, these long-term investments will mean more pressure for Intel's margins, but this spending is necessary to remain a player in the chip industry.

From the beginning Intel knew the game was about managing product development cycles

Gordon Moore, Intel's founder, predicted that the number of transistors on a chip would double every two years. Moore's law was put into practice for many years and it ultimately led to Intel's tick tock strategy. Intel's 'Tick Tock' idea was developed in 2007 and refers to the change in its processor technology. Each year, the ticks and tocks alternate, ticks refer to downsizing of the previous micro-architecture and tocks refers to new micro-architecture. During a tick cycle, new process technology is developed, which enhances the performance of previously produced microprocessors. A tock cycle consists of an entirely new processor or product. Many say that this rhythm of development sets Intel apart from other companies, as having a fixed timeline allows the business to produce products at pre-determined intervals, and yet manage the product profitability carefully. 'Intel has successfully alternated and delivered the next generation of silicon technology as well as new processor micro-architecture year after year' (podtech.net) [2]. Eventually, competitors caught on to this lucrative market. In 2006, Intel created Core micro-architecture, considered a 'tick' in the 'tick tock' model.

In the beginning Intel produced memory chips

In the early 1970s, Intel produced SRAM, ROM, and some transistors. These memory devices allowed Intel to dominate the market early on, as the company increased production and expanded its marketing horizons. A few years later, the emergence of memory chips became prevalent and the competition intensified.

In a bold move, Intel decided to focus on microprocessors, hopefully cornering a new target market. Becoming a supplier for IBM, Intel's business (and profits) grew exponentially. Intel's shift to microprocessors was rather lucky. It was able to invest in the microprocessor market quickly, leaving little time between investments and, in this case, extremely positive outcomes. This gave Intel a competitive advantage in the market.

Intel also knew how to protect its turf

Intel maintained 'proprietary standard', as it was IBM's sole supplier, suing anyone who dared to imitate its product, making it nearly impossible for other companies to enter the market. Intel was also able to create brand awareness through campaigns such as 'Intel Inside.' Successfully acquiring IBM and Apple as clients allowed Intel to maintain market share. The company was able to change the industry dynamic, allowing it to release new products every year while reinvesting in its old development.

Transition from memory chips to micro processors changed the company

During the transition from memory to microprocessors, Intel's number one concern was becoming the best microprocessor company. Former CEO Andy Grove created a system called Strategic Long Range Planning. Essentially, this strategic framework allowed Grove to do things his way. 'I became very distinctive in prescribing the strategic direction from the top down. This defined the strategy for all of the groups, and it provided a strategic framework for different groups at different levels of management.' (Grove as quoted on aomonline.org).[3]

His strategy made Intel a more focused organisation. The company created a matrix, consisting of both product and functional groups. Product groups consisted of things such as desktop products, while functional groups consisted of sales and marketing teams, among others. Working with companies such as IBM and Microsoft led to much of Intel's success, as the majority of Intel's products were used exclusively in the products from these companies in the 1990s. Looking forward, Intel desired to reach a new target market by working with the software industry.

Sales volume feeds ability to invest in product development which feeds more sales

As stated earlier, with intent to dominate its markets, Intel's goals were, and still are, to release as many microprocessors as possible in the shortest amount of time, allowing them to maintain market share and invest in new technology. To achieve such ambitious production volumes, Intel spends billions of dollars annually on capital investments, primarily to maintain and/or increase production capacity.

However, looking at economies of scale, we see that 'volume is key to success in this game. A leading edge design will take 50-100 top engineers two to three years to develop.' (aomonline.com) [3] Capital investments, when developing a new product, can cost up to $800 million. Selling in large volume helps Intel off-set the production costs.

Intel is also continually refining its market understanding

Intel has started targeting a new market segment: 19-24 year olds, who generally have a large influence within the marketplace. With the strongest quarterly results in its history, Intel is determined to maintain this positive streak by targeting new consumers. Deborah Conrad, Intel's chief marketing officer, says '[computers] are an indispensible part of [Generation Y] lives. For that reason we need to be relevant to them... they don't want anything that isn't authentic. So we're creating something that is' (Forbes)[4]. She goes on to mention 'buying a computer is difficult; therefore, we need to get people to care about our brand, that our technology is more than just computer technology. Intel is determined to make products that appeal to this market by making them more exotic and exciting.' [4]

Intel has always focused on research, development and production of its products

Intel uses vertical integration to help it stay ahead of competitors. It does not outsource any of its production;

rather, it does it all in-house, spending close to $7 billion last year on manufacturing plants. While this is an expensive endeavor, Intel claims it ensures quality, protects its ideas, and can meet its tight time schedules, giving it a competitive advantage. Intel invests large sums of money in high quality, clean production, which not only sets it apart from the competition, but also allows the company to produce a quality product in large volume and get it to the consumer quickly.

Supply chain plays a crucial role at Intel

Intel's supply chain contributes hugely to its success. Intel understands that in order to release new products while investing in previous developments, it must have a supply chain that works flawlessly and efficiently. Supply chain advantage is another reason why Intel has been able to invest in its own developments as well as new technology. 'The importance of the supply chain in Intel's business growth has steadily increased over the last decade' (Intel IT)[5].

One of Intel's strong points is the size of markets it serves. Having such a diverse group of products, ranging from wireless controllers to software, has allowed Intel to win in the marketplace. However, such a large range has multiple supply chain requirements. In the global marketplace, expanding products to new markets and shrinking product life cycles has forced force Intel to adapt to new supply chain requirements and processes in order to bring a product to the consumer effectively.

Intel works with its 5-STAR Business Network of collaborators to enable it to launch newer products

One key concept in the microprocessor business is that each company's ideas build off of one another. This has led to a collaborative approach between companies. While Intel and AMD are rather distinctive competitors, companies such as Microsoft and IBM have helped facilitate Intel's success.

Intel collaborated with Microsoft in the 1990s, contributing to much of Intel's success. This success has led Intel to become a PC industry leader. Tom Yuen, co-founder of PC maker AST Research, says 'You no longer buy a Compaq computer, you buy an Intel computer from Compaq' (aomonline.org).[3]

Figure 11.9 on the next page represents the crux of the argument in this case. The microprocessor is but one element of the overall customer experience with the personal computer. Other elements - such as motherboards, memory, operating systems, application software and other ancillary products - must be developed in step with the micro-processors to enable a streamlined customer experience. Imagine an APP diagram similar to the one shown in Figure 11.8 (page 226) for Intel, for each of these other players in the ecosystem. Those players who did not keep up with the game would eventually fall by the wayside, only to be replaced by more adept players.

Intel had (and still quite possibly does) the ability to influence the balance of power within the chip industry. 'In the 1990s, Intel was a major player in the market for chip-

Figure 11.9: Other companies developed products to enable Intel's APP and Intel did the same for them

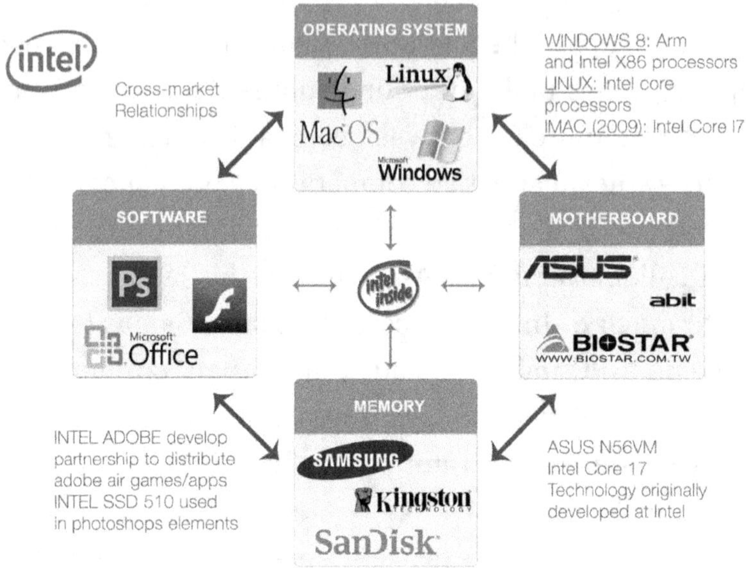

sets and motherboards. This allowed Intel to support OEM manufacturers who were less competitive, leveling the playing field among OEM manufacturers somewhat and it offset some of the power and competitive advantage of the major OEM manufacturers' (aomonline.org).[3]

Overall, Intel appears to be doing the impossible: Investing in new technology while simultaneously maintaining and investing in its existing product line-up developments. This system creates a pipeline of continuously flowing products, which are both innovative and sold in large volume, making Intel one of the most successful Fortune 500 companies while producing the largest amount of microprocessors.

SUMMARY

Success comes only when the innovation cycle continues in perpetual motion. Effort must be simultaneously focused on the current product, as well as future products. Supply chains cycle between adaptive and integrative as companies switch from one generation to next. It is a balancing act - between the current needs for profits and future needs for products. It is unprofitable to launch products too frequently. It is suicidal to milk products beyond their useful life. 5-STAR Business Networks help take product phasing to a new level. Intel has been executing APP very successfully for a very long time. From the beginning, Intel knew the game was about managing product development cycles. Intel knew how to protect its turf. Sales volume feeds ability to invest in product development that feeds more sales. Intel is also continually refining its market understanding. Supply chain plays a crucial role at Intel. Intel works with its 5-STAR Business Network of collaborators to enable it to launch newer products.

RESULTS-FOCUSED OUTSOURCING AND MODULARISATION

Chapter 12

Outsmart, Outsource, Outprofit

> ***"And stand together, yet not too near together: For the pillars of the temple stand apart, And the oak tree and the cypress grow not in each other's shadow."***

Khalil Gibran, The Prophet

When I studied at a maritime academy, navigators who specialized in dynamically positioned navigation were a highly envied lot. Not only did they get the best training - to use multiple propellers on the vessel, ship, oil rig or such to keep it virtually stationary, counteracting all type of forces - but the pay and conditions were excellent. Deepwater Horizon, a semi-submersible oil rig owned by Transocean was a dynamically positioned drill rig. Highly acclaimed for its many deepwater successes, the oil rig was deployed off the coast of Louisiana at an approximate cost of $1 million per day to drill an exploratory well for British Petroleum (BP), which held the exploratory rights for the block it jointly owned with two other parties. BP had chartered the oil rig from its owners, the Swiss entity, Transocean. Although Deepwater carried a Marshall Islands flag, Transocean operated the rig through a subsidiary, Triton Asset Leasing, also based in Switzerland. We are examining this long case study in order to garner some relevant lessons regarding outsourcing.

5-STAR Business Networks come together in many forms

Continuing with the case study, as the exploratory well neared completion, on April 20, 2010, Deepwater Horizon became front page news in nearly every newspaper on earth. The incident was reported in a press release by Transocean [1] as follows:

> *"Transocean Ltd. (NYSE: RIG) (SIX: RIGN) today reported a fire onboard its semisubmersible drilling rig Deepwater Horizon. The incident occurred April 20, 2010 at approximately 10:00 p.m. central time in the United States Gulf of Mexico. The rig was located approximately 41 miles offshore Louisiana on Mississippi Canyon block 252."*
>
> *"Transocean's Emergency and Family Response Teams are working with the U.S. Coast Guard and lease operator BP Exploration & Production, Inc. to*

> *care for all rig personnel and search for missing rig personnel. A substantial majority of the 126 member crew is safe but some crew members remain unaccounted for at this time. Injured personnel are receiving medical treatment as necessary. The names and hometowns of injured persons are being withheld until family members can be notified."*

The details of the incident, as per the figures from Popular Mechanics [2] were attention grabbing:

> ***4.9 million**: Barrels of oil (205.8 million gallons) leaked from the Deepwater Horizon well, about half the amount of crude oil the U.S. imports per day*
> ***19**: Times more oil leaked from Deepwater Horizon than spilled from the Exxon Valdez in 1989 (10.8 million gallons)*
> ***62,000**: Barrels leaking per day when the wellhead first broke, roughly the amount of oil consumed in Delaware each day*
> ***53,000**: Barrels leaking per day when the well was capped on July 15, roughly the amount of oil consumed in Rhode Island each day*
> ***397.7 million**: Dollars' worth of the oil spilled at current market prices ($81.17 per barrel)*
> ***665**: Miles of coastline contaminated by oil*

The resulting investigation to establish the causality, contributing factors and liability will fill up a book many times the size of the one you are holding.

Outsourcing is a fact of life in business today

We will, however, briefly focus on the three relevant parties - BP, Transocean and Halliburton - for the sake of discussion relevant to this chapter, modularised outsourcing. BP had outsourced the task of drilling to Transocean. At the same time Transocean had bought the Blowout preventer from Cameron International Corporation. Whether it can be argued that BP or Transocean had outsourced the task of Blow-out Prevention (BOP) to Cameron, it is not certain; neither is the liability on malfunction of the blowout preventer because of allegations of lack of proper maintenance. Cameron agreed to settle all claims related with the Deepwater Horizon tragedy with BP for $250M - without any admission of guilt. The situation with Halliburton is still unclear. As per a CNN news report[3]:

> *BP and Halliburton sued each other in April 2011 claiming each is to blame for the deadly explosion on the Deepwater Horizon rig and resulting disastrous oil leak. Halliburton was in charge of cementing the*
> *Macondo well and claims that its contract with BP indemnifies (releases) Halliburton of any legal action resulting from its work as a contractor...*
>
> *In a response filed Sunday, BP asserted that "maritime law prohibits indemnification for gross negli-*

gence."

As part of that four-page filing, BP reiterated that it was seeking to recover from Halliburton "the amount of costs and expenses incurred by BP to clean up and remediate the oil spill." BP has estimated in the past that the total cost will be around $42 billion, and by the end of November 2011 the oil company it has paid out or agreed to pay out $21.7 billion to affected individuals, companies and governments around the Gulf.

In an e-mail to CNN, Halliburton spokesperson Beverly Stafford said "Halliburton stands firm that we are indemnified by BP against losses resulting from the Macondo incident."

Outsourcing tasks does not transfer responsibility for those duties

US President Barack Obama quipped in an interview with CNN[4] On May 14, 2010 "you had executives of BP and Transocean and Halliburton falling over each other to point the finger of blame at somebody else... The American people could not have been impressed with that display, and I certainly wasn't." The legal wrangling continues and will take considerable time and expense to resolve. We need not go into the gory details of dollar numbers too big to even fully comprehend, but from our perspective in this chapter, three key useful insights stand out:

1. When you outsource a task, service or suite of services, you could still retain significant responsibility for its full and proper execution.
2. Brand-names, size of the company or even their experience is no guarantee of performance of the outsourcing service contracts.
3. No matter how close the 'partnership' is at the start, the test of a successful outsourcing contract is how it ends.

There hardly exists a corporation today which does not outsource at least part of its tasks to third parties. The trend towards outsourcing is only accelerating as businesses discover the benefits of outsourcing - speed, flexibility, accountability and ability to focus on core competence. **Super networked businesses outsource even more often than other businesses - yet they succeed more often where others fail relentlessly. How do they do it?**

Outsourcing has a long history

The advent of outsourcing is best described by Paul Anthony McMahon in his doctoral thesis at Southern Cross University[5]:

> *Outsourcing, or as it was then labeled "contracting out", has been in use on an industrial or commercial scale since the advent of the industrial revolution in*

England during the 1700's (Brown & Wilson 2005; Kakabadse & Kakabadse 2003), with firms facing the "make or buy" conundrum that resulted from the greater production efficiencies that characterized eighteenth century England (Domberger & Hall 1995). In support of this contention, Greaver (1999, p.10) wrote that "outsourcing is similar to subcontracting, joint venturing, and strategic partnering concepts, which date back hundreds of years", citing the following examples: farmers hiring migrant workers; construction companies subcontracting electrical and plumbing activities; and governments subcontracting defence material production to private companies.

The first systematic use of outsourcing can be traced back to the 1940s, during World War II, when organisations provided systems facilities management services to the US government (Greaver 1999). However, it was the growing dissatisfaction with the underperforming post-World War II ideal of economy-of-scale driven conglomeration (Hunter & Cooksey 2004) and the introduction of timesharing mainframe computer services in the 1950s and 1960s (Factor 2002) that set the scene for the wider adoption of outsourcing methodologies.

So, what are today's outsourcing contracts about? Currently, you can outsource almost any service right from IT to logistics to accounts payables processing.

Janitorial services are the most commonly outsourced

Despite the hype that customer service is the most commonly outsourced service, that honour belongs to janitorial service. When they discuss outsourcing, business process outsourcing (BPO) mavens also almost completely forget about shipping, transportation, warehousing, logistics, contract

Figure 12.1: Organisational Activities

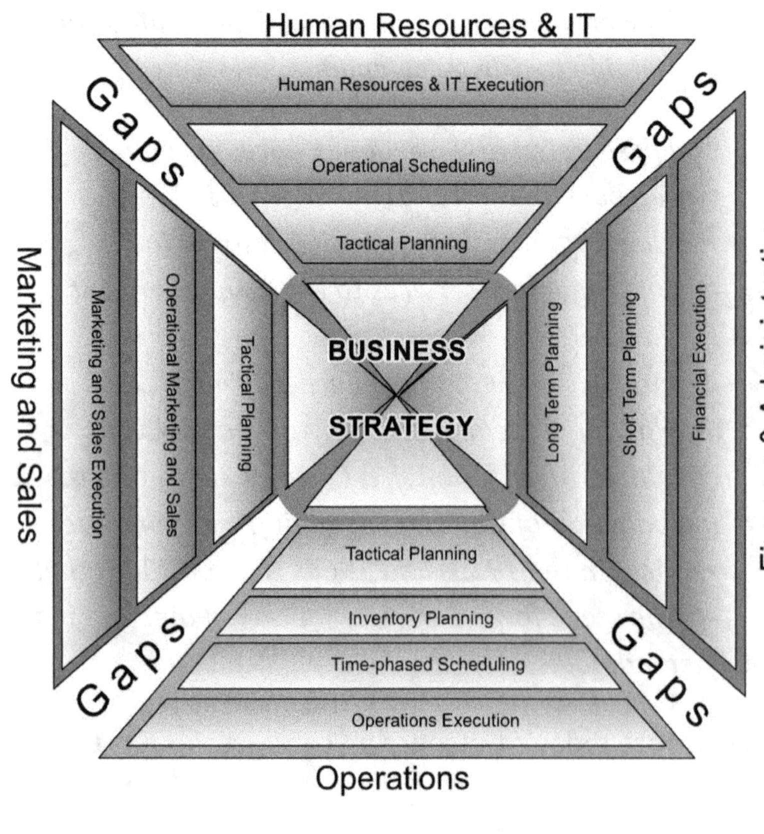

manufacturing and similar services.

The full gamut of organisational activities are shown in Figure 12.1.

The four key functional areas at four escalating levels of commitment to the core make up a 4x4 matrix of most of the activities carried out within an organisation. Except for the top layer in each pyramid (covered with in red square), which is largely strategic in nature and highly sensitive to an organisation's success, almost all other activities are open to outsourcing under the right circumstances.

Each of the four key pillars constitutes services that can be outsourced

As we can see in Figure 12.1, each of the four key pillars of the organisation works in its own silo, although often there are gaps between them. The gap between the activities of the sales and marketing department on one side and the operations and logistics department on the other, needs to be filled in with suitable organisational mechanisms. In this case, sales and operations planning (S&OP) is frequently the right mechanism to fill the gap between the activities of the sales people and the operations staff.

Similarly, the gaps between the activities of the finance/ accounting department and the operations department is often handled by utilisation of cost of service (CoS) methodology. In fact the core principle of the Supply Chain Management discipline is exactly that - to bridge the gap between organisational silos so that they act as a single organisation; S&OP and CoS are just two of the key mechanisms that achieve cohesion.

Creative strategists keep coming up with more and more services that can be outsourced

The most popular activities that are outsourced in each of the functional areas are shown in Figure 12.2 on the next page.

Figure 12.2 is by no means a full and exhaustive list of every organisational activity that can be outsourced; companies continue to find more and more activities that can be more efficiently and effectively done by entities outside the company.

The trend towards outsourcing continues to accelerate

Despite growing emphasis on outsourcing, the results are mixed. A significant proportion of outsourcing contracts continue to fail decades after they first were put in place. Survey after survey confirms a vast majority of outsourcing deals end in unpleasant divorces, if not outright hostilities. Yet, the trend towards outsourcing shows no sign of abating. So, the demand for outsourcing services is growing while suitable supply continues to be constricted because of its inability or unwillingness to understand the demand properly.

Objective research on outsourcing is difficult to come by in this emotionally charged area. Job losses, off-shoring, declining skill base and growing income disparity levels all combine to create an atmosphere of paranoia. Yet even the

Figure 12.2: Organisational Activities and Outsourcing

Information Technology

- Systems Integration
- Software selection
- Software configuration and customisation
- Hardware supply, maintenance and repair
- User training
- Applications development
- Consulting and reengineering
- Mainframe data centers - operations and maintenance
- Networks - configuration, operations and maintenance
- Complete Information Technology outsourcing
- Information security

Human Resources

- Recruitment
- On-boarding
- Training
- Performance Management
- Goal setting and
- Succession Planning
- Relocation
- workers compensation
- Human resource information systems
- Payroll processing
- HR policies and procedures evaluation and formulation
- Talent management planning and execution
- Company culture review

Marketing

- Advertising strategy
- Market Research
- Promotional materials
- Pricing Strategy
- Advertising Buying
- Loyalty programs - creation, operation and management.
- Direct Marketing Campaigns
- Tele-marketing campaigns
- Marketing call centre operations - inbound and outbound
- Conference and workshops - planning and operations

Sales

- Direct sales campaigns
- Sales force Training
- Prospecting
- CRM systems - configuration, operation and management
- Reservations and sales operations
- Field sales and service operations - planning and execution
- Sales force effectiveness studies
- E-commece selling
- Sales teams incentives planning and delivery

Operations

- Contract Manufacturing
- Production Planning and scheduling
- Procurement Management
- Strategic Sourcing
- Product design
- MRO for production plants
- Production plant debottle-necking
- Statistical Process Control (SPC)
- Inventory Management
- Operations optimisation systems - configuration, installation and upkeep
- Security of premises

Logistics

- Contract Logistics
- Transportation
- Warehousing
- Value added warehousing
- Transportation planning
- Freight Audits
- Transportation brokerage
- Fourth Party Logistics - Management of outsourced logistics function
- Shipping
- Ship-brokering
- Freight rate negotiations
- Transport equipment certification
- Insurance in transit

Finance

- Capital Raising
- Debt and Equity management
- Hedging of currency and time risks - Financial risk management
- Investors and market communications
- Capital control
- Capital rationing
- Project financing
- Trade financing
- Treasury management
- Security of cash and cash instruments
- Value-at-risk management

Accounting

- Financial Statements auditing
- Financial statements preparation
- Budgeting
- Budget variance analysis and corrective actions
- Taxation strategy and planning
- Taxation strategy execution
- Asset Valuation and Revaluation
- Asset register maintenance and safekeeping
- TOP

most vocal protesters seem to enjoy the fruits of the trend and continue to buy the latest smart-phone or flat screen TV. This leads me to believe that the trend will not only continue but also gradually accelerate - it is impossible to put the genie back in the bottle.

Despite its long history and huge growth, outsourcing continues to provide unacceptable results

Let us examine the evidence on satisfaction with outsourcing. Commonly accepted and relatively unbiased opinions seem to be generally agreed upon:

1. About one-quarter of outsourcing contracts fail within one year
2. Nearly up to 70 percent of outsourcing contracts are regarded as failures by one of the two parties
3. Rigorous testing - such as meeting initial objectives - might show an even higher proportion of contracts would be considered as failures.

But what are the common reasons for this? Again, the answer varies from survey to survey, and rather than quote one survey or report, we summarises the common findings below:

- **a.** Poor strategic thinking and lack of preparation to understand the key motivation for outsourcing
- **b.** Lack of sufficient depth of thinking upfront is perhaps the single biggest cause of failure. If there is no com-

mon and mutually agreed vision of how it is going to end, it generally ends badly and quickly.

c. Poor governance and communication through the project is also accepted as a big culprit. There can be a variety of reasons for this, including point 'a.' above.

d. Information asymmetry, selective disclosures or outright low-balling, over representing capabilities, over-promising and under-delivering, cost creep, and mis-aligned incentives, make up the rest of the reasons frequently quoted for failure of outsourcing deals.

However, it must be recognised, that a vast majority of small tasks that are outsourced continue to be successful and create happy, long-term customers.

It appears that complexity or size of task increase the probability of failure of outsourcing

While the research on the size of outsourcing contracts and its relationship with the success of outsourcing relationship is currently sparse, it appears to be the case that majority of failures referred to above were in situations where turnkey solutions were outsourced to entities that planned, executed, supervised, measured and reported on a full gamut of tasks in a particular area - e.g. in logistics or in information technology. The problem appears to result from lack of willingness to carry out the task because of lack of the separation of powers between planning and execution, rather than with the ability to carry out the tasks. After all, if the entity that plans also executes, it will plan it in such a way that suits itself rather than the customer - especially if it enjoys information or management advantage.

Two key approaches to fix the problem

Faced with this problem, two approaches have been tested. One pre-dominantly in the USA and the other mostly in Europe - both with partial success.

US-centric approach

The US-centric approach takes a legalistic view of very detailed contractual relationships, resulting in micro management of the outsourced relationship through service level agreements (SLAs). The contracts run into several hundreds, sometimes thousands, of pages. The attempt is to think of every eventuality that might arise during the currency of the contract and to allow for it. The contracts frequently allow even for that which is unforeseeable. Government and public sector outsourcing frequently follow this model also. In our experience, though, outcomes from such outsourcing arrangements are no better than those through simple contracts with a limited number of key performance metrics and virtually no SLAs. While they give a false comfort of contractual coverage before outsourcing, most business situations are so dynamic and the mutual interaction between the various parties is frequently so intertwined that no SLA can clearly define it, let alone manage it. Faults and shortcomings are frequently difficult to pinpoint and the party that carries the balance of power ends up carrying the arguments in most cases.

Euro-centric approach

The second approach is to make the outsourcing service provider responsible for the outcomes and clearly relates its payoff with the results of the service. There are several variations of this approach and many different names are applied to it - including Vested Outsourcing, Results-based Outsourcing etc. It has a more successful track record in our experience. The reasons for the success are obvious, explicit link between the results and payoff of the service provider. There are, however, still some issues. Firstly, you only know the results after the fact and, by then it might be too late as your company is already behind the competitors in the game. Secondly, the problem with separation of powers between planning and execution still remains in place. In situations where the key problem pertains to the inherent conflict of interest between the results of planning and results of execution, if both of these functions are carried out by the same party, results tend to be lackluster.

Conflict of interest between the service provider and the customer

If planning shows cost reductions that will result in revenue shrinkage for the service provider, it will be a rare service provider who will carry out both the planning and execution in such a manner where their own revenue shrinks significantly. Rarely is there a musician who can play an instrument and conduct the orchestra at the same time.

The solution, then, must go past a results-based or vested outsourcing approach. The problem is big enough for corporations to spend hundreds of millions (or even billions) of dollars every year on litigation as well as contract creation. The costs of failed contracts, liability, failure to provide services or goods on time and other downstream secondary effects cannot even be measured in many cases (the oil rig case mentioned at the beginning of this chapter is a case in point.)

On one hand, it appears cavalier to leave most of the details of execution of the contract to the service providers until its too late. On the other hand, the rigidity and inflexibility of a micro-managed contract leaves no room for initiative, growth or flexibility of business strategy in the face of changing circumstances.

What, then, is the best way forward?

5-STAR Business Networks use Results-focused Outsourcing and Modularisation (ROM) very effectively

A successful approach we have used combines results-focused outsourcing with modularisation networks. So what is modularisation?

Let me illustrate the magic of modularisation by using the case study of standard shipping containers instead of boxes and crates.

A similar move towards modularisation of global business networks is emerging and will have a far-reaching impact on

Case Study - Containerisation of the global shipping trade

The field of international goods transportation has witnessed several great advances over the last 100 years. Of these, perhaps, none had more far reaching impact than the advent of containerisation. Before 1965, all packaged goods were packed into boxes of different sizes and shapes to be loaded on ships. General cargo ships themselves were small in size – no bigger than 20,000 tonnes or so – and stayed anywhere between 14 weeks in any single port. About 70 percent of a ship's time was spent in non-value producing activities such as planning and loading boxes and crates of diverse sizes and shapes into the ships holds.

Containerisation changed all that. Ships are now built to carry standard containers of 20' (or 40'). Containers are packed and unpacked in container yards far from ports – reducing the need for holding the ships in port, and costly warehousing space close to port. Size of container ships has increased greatly. The latest ones will carry nearly 10,000 TEUs (twenty equivalent units) or about 100,000 tonnes. These ships utilise more than 80 percent of their time on value-producing activities. No wonder the cost of shipping one cubic meter of packaged cargo has gone down by nearly 75-85 percent in real terms over the past 40 years. This has made it possible to reconfigure the global supply chain in such a way that most activities are now carried out in the best place to do so. The manufacturing boom in China, super large container shipping companies, the hub and spoke model of shipping and the emergence of 6-10 global container terminals, are all partially a result of containerisation.

the organisations and businesses.

Imagine a business network which is able to instantly flex itself to the changes in your business structure and dynamics. Every time your company bought a division, you could seamlessly integrate its parts with your overall business. Every time you divested a business, you could unravel the integrated business structure without too much pain (and/or expense). Moreover, every time a manufacturing, design, assembly, trucking, freight forwarding, warehousing or shipping company proved itself incapable of meeting commitments made during the sales process - you could easily take your business to another one.

By trial and error many businesses have moved towards modularisation of their entire networks

Every time you reconfigure your manufacturing footprint, buy or sell a business division, change your business strategy or alter your sourcing arrangement, your business network follows closely. In other words, your business is not beholden to any one service provider or systems provider - because they are all interchangeable. The physical infrastructure, processes, systems and services involved in configuration of global business networks are slowly, but inevitably, being modularised by the business leaders.

Case Study - 1960s - Modularisation of the global ship building industry

During the 1960s, the shipbuilding industry reduced the cycle time for building a large-ocean-going merchant vessel from three years to six months by modularisation. The whole ship's structure was divided and sub-divided into smaller and smaller modules that were produced simultaneously by a number of satellite suppliers just in time to be brought together and joined together into a final assembly. The whole system is still one of the finest examples of a well-orchestrated supply chain in motion. Tremendous benefits in terms of cost and production time reduction, increased throughput, flexibility and customisation ability accrued to the shipping industry as a result.

The Japanese ship building companies utilised this emerging trend to their great advantage. With their Kieretsu-type structure and lean manufacturing processes (and government assistance), they rapidly built substantial cost advantage and market share that lasted until the South Korean shipyards (and now Chinese shipyards) copied them. A typical Japanese shipyard was surrounded by ancillary businesses that supplied modules and sub-modules to the shipyard and were totally dependent for their business on the parent company. Close co-ordination eliminated waste, substantially reduced cycle times and led to significantly lower turnaround times.

Systematic approach to modularisation and outsourcing yield better results faster

Figure 12.3: Modularisation - a systematic approach

Great companies outsource strategically —they modularise to commoditise.

While the service providers and suppliers like to build unassailable relationships via bundling and high switching costs, great companies recognise that they must keep control of their operations at all times. Not just appearance of control - but real control. Perils of losing control include losing competitive advantage over time, cost creep and eroding margins, nasty surprises, information asymmetry and market share loss.

Great companies unbundle the offerings of service providers and choose real value where ever they find it to assemble their own supply chain to match their business strategic thrust. They modularise the supply chain into small interchangeable parts and use service providers discerningly.

STANDARDISATION
(emergence of common parameters-terminology, measurements, activities)

↓

MODULARISATION
(breaking up into smaller more manageable units)

↓

HOMOGENISATION
(creation of similarity between modules in order to facilitate switching)

↓

COMMODITISATION
(multiple buyers competing on a level field to supply modules at best price)

Modularisation is accompanied by Standardisation and Homogenisation

While a detailed description of modularisation will be the basis for my next book, Figure 12.3 indicates in brief the four key elements involved in the process. The example of the PC industry is a case in point.

Modularisation, in this context, means the breaking up of a greater whole into interchangeable parts that fit together seamlessly, and together, in many different combinations and permutations, make many different wholes. The gradual evolution towards modularisation of business networks is being spearheaded by the customers who demand control and flexibility in their businesses, coupled with low cost.

As shown in the graphic, modularisation is preceded by standardisation – achieving commonality in terminology, in measurements and in activities. Standardisation is essential to modularisation because it helps simplify the complexity and bring uniformity. Imagine, if shipping containers were of many different sizes. Efforts to modularise general cargo shipping would have been seriously hampered.

Why is modularisation attractive? Because it helps us achieve homogeneity – a condition where products and services start looking more and more similar. Homogeneity, in turn, encourages substitution and/or switching between suppliers and hence commoditisation of the relevant market. This essentially leads to falling prices and growing volumes, both leading to massive growth in business and commerce, and improved standards of living.

Case Study – 1980s - Modularisation in the PC industry

Perhaps the clearest example of modularisation is the evolution of the personal computer industry in the last 25-30 years. 1981 saw the launch of the IBM Personal Computer. The IBM PC brought together all of the most desirable features of a computer into one small machine. It offered 16 kb of user memory (expandable to 256 kb), one or two floppy disks drives and an optional color monitor. When designing the PC, IBM for the first time contracted the production of its components to outside companies. The processor chip was from Intel and the operating system from Microsoft. IBM's decision to outsource the two key components eventually led to the modularisation of the computer industry.

Dell utilised the emerging trend of modularisation in the PC industry to create a unique business advantage. While it offers highly customised computers to its customers, the whole production chain and final assembly are highly standardised. This is due to the way a computer is built. All the components and sub-components such as the motherboard, hard disk, graphics cards, memory etc. are available in a huge variety and are essentially interchangeable modules performing essentially similar tasks. This makes it easy to offer the technical specifications that the customer wants. This and the fact that the computer chassis is easy to manufacture in different forms and colours and with the help of plastic "appearance parts" give it a wide variety of looks; and makes it easy to offer a completely customised computer to customers.

Results-focused Outsourcing and Modularisation (ROM) puts responsibility where it belongs

One of the most persistent myths that continue to be perpetuated despite ample evidence to the contrary, is once you outsource you can relax and let them handle all the work. This is perhaps the most pervasive myth in the higher echelons of corporate circles today. Even some of the most astute business executives tend to believe the hype that once they outsource any function (generally considered non-core), they can stop worrying about the outsourced functions. However in reality, this can be dangerous. Putting together an outsourcing arrangement does not transfer your core issues to the service provider; rather you swap one set of problems with another, e.g. internal logistics management problems with logistics relationship management problems!

While this topic alone calls for an entire chapter, let's look at some of the problems likely to surface after outsourcing begins: Unpleasant surprises - such as cost creep, technical expertise being lower than expected, lack of flexibility to incorporate changing business needs, potentially large cost overruns on account of services that were assumed to be included in the service contract but were not, loss of real visibility, control and direction - are just some of the issues often seen in this context.

The fact is that even after the outsourcing, your management will still be working just as hard on the outsourced functions, the sole difference being that the time they save on the management of the outsourced functions will now be spent in managing the relationship, with some relationships even degenerating into games of contractual head-butting and variations management for which both the IT outsourcing and the building construction industry are becoming notorious.

To achieve proper power balance and to facilitate easy switching from underperforming service providers, modularisation of processes is required. In order to manage the contract properly, a good mix of details oriented contracts and on-going relationship management are required.

Results-focused Outsourcing and Modularisation (ROM) makes sure that the customer retains control of the relationship

Another pervasive myth – often perpetuated by the organisations handling outsourced functions - is that of an equal win-win partnership. A real partnership, by its very nature, is defined by co-dependence where the relationship is indispensable to both parties, as judged by a measure of how much each party would be hurt if the relationship fell apart. The truth, in most cases of outsourcing, is that this pain is not too much!

Most service providers have fingers in multiple pies, and have learnt not to rely on just a few customers for their business, for very obvious reasons. From their point of view, even if a few customers pulled out, they would have contingency plans to optimise their asset utilisation. Even from the customer's perspective, with dual sourcing and the increasing modularisation of the supply networks, most service providers are now providing a commoditised service, despite all the hype surrounding technology integration, business process integration and infrastructure integration. None of the service providers are considered indispensable. And this is true of all service providers regardless of their scale of operations.

Even if it were a true win-win partnership - as when a couple dances one person takes the lead to improve joint performance - it is preferable that the lead be taken by the customer whose business is at stake. Corporations should relinquish the lead to the contractor only at their own peril.

Results-focused Outsourcing and Modularisation (ROM) is an intricate dance which commences well ahead of the actual outsourcing

The approach that we have developed and successfully applied combines the key parameters discussed above in order to effectively prepare, engage, select, integrate and manage the service providers for the modularised set of services. In fact, those five words are the five key steps in the process as shown in Figure 12.4 (page 263). Each step consists of further sub-steps - each one of which has its own set of templates, heuristics or methodology to facilitate rapid deployment of ROM strategies.

Obviously the details of this approach are beyond the scope of this book. But the results achieved through a systematic, logical and pragmatic approach such as this one far surpass any others. Following are some of the key lessons:

Lessons from the leaders in 5-STAR business network modularisation:

1. Business network strategies are increasingly important – something which could never be outsourced to your service provider or a contract manufacturing outsourcer.
2. Great companies are marvelously adept in their ability to unravel the packaged bundles of services presented to them, pick and choose the components that suit them the best, and put together a best-of-breed customised bundle for themselves.
3. Modularisation makes it easier to homogenise and hence commoditise a market. Hence, whether it is the market for systems, for processes, for physical infrastructure or for services, these are increasingly commoditised going forward. Companies will do well to prepare for this coming commoditisation and stay on the leading edge of the curve.
4. Modularisation has changed the face of global businesses. All tasks are carried out where it is most cost effective to do so. You can expect service factories to be the norm, working in unison with physical factories in different locations.

Figure 12.4: ROM

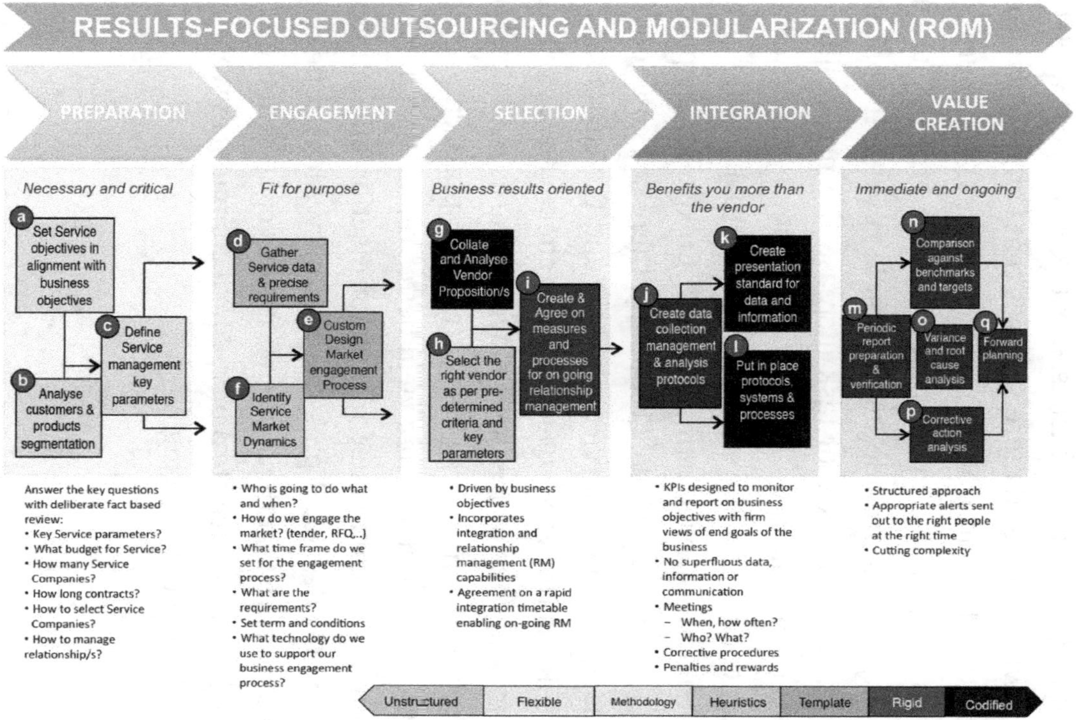

SUMMARY

Outsourcing has a long history and outsourcing is a fact of life in business today. Outsourcing tasks does not transfer responsibility for those tasks. Creative strategists keep coming up with more and more services that can be outsourced, so the trend towards outsourcing continues to accelerate. However, despite its long history and huge growth, outsourcing continues to provide unacceptable results. It appears that complexity or size of task increase the probability of failure of outsourcing. Therefore, two key approaches are being tried to fix the problem: US-approach relies primarily on micro-management, whereas Euro-centric approach is macro results driven. Frequently, there is a conflict of interest between the service provider and the customer. 5-STAR Business Networks use ROM very effectively. By trial and error, many businesses have moved towards modularisation of their entire networks but systematic approach to modularisation and outsourcing yield better results faster. Modularisation is accompanied by standardisation and homogenisation. Besides, ROM puts responsibility where it belongs and makes sure that the customer retains control of the relationship. It is an intricate dance, which commences well ahead of the actual outsourcing.

Practical Example of 5-STAR BUSINESS NETWORKS

SECTION 3

PRACTICAL EXAMPLE OF 5-STAR BUSINESS NETWORKS

> ***"He who loves practice without theory is like the sailor who boards ship without a rudder and compass and never knows where he may cast."***
>
> *Leonardo da Vinci*

> ***"An ounce of action is worth a ton of theory."***
>
> *Ralph Waldo Emerson*

In the first section of this book we examined the networks in general. We examined what the networks are, their key characteristics and benefits, their history and the impact of technology on the networks.

In the second section we looked at the business networks in their full glory. We examined how the modern supply networks are configured for global leadership and how they enhance the profitability of the companies that participate in these networks. We examined the key cornerstones of the 5-STAR Business Networks and how they form the underpinnings of success in modern commerce.

In this last section, we have only two chapters which examine in considerable detail a 5-STAR Business Network par excellence. The intent of Chapter 13 is to see all the concepts in action at once and discern how they can be adapted and adopted in your own business.

Chapter 13 - Super networked business of the third-richest man on Earth

- Create your own business model: never copy the business network or model of your competitors, suppliers or customers. Authenticity, originality, fit-for-purpose business models win the day every time.
- Keep fine tuning your business model and the 5-STAR network continually as the circumstances change. Customer preferences evolve, newer suppliers become available, newer technologies come to the market and competitors change their strategies. All of these will require changes to your business model and the 5-STAR network.
- Whenever there is a big change in your business - merger, acquisition, divestment, expansion, new market entry - re-consider the appropriateness of your business model and the 5-STAR network. What was appropriate

until yesterday may not be appropriate anymore.

- Know your customers: the more you know your customer, the faster you can translate that knowledge into products they are willing and able to pay for, the higher your ability to win in the marketplace. No matter how big your company becomes, do not lose touch with your customers. This customer knowledge is the primary source of innovation - keep it in front of everything your company does.
- Make your suppliers dependent on your business: become one of their biggest and best customers if you can. That way you can create a multitude of smaller suppliers - all part of your 5-STAR networks who will play the game as per your business strategy.
- Use technology to its best effect: keep investing in technology to gain maximum advantage towards your business strategy.
- Create a buzz in the marketplace around your products, around your brand, and around your offerings. Keep the marketplace buzzing as much as possible via customer networks and conversations.
- Outsource strategically and modularise before outsourcing. Do not outsource to any single entity. Make outsourcing work for your business, now and in future.

Chapter 14 - Research methodology and other companies with 5-STAR Business Networks

- Research methodology
- Measures of network efficacy
- Other companies with 5-STAR Business Networks

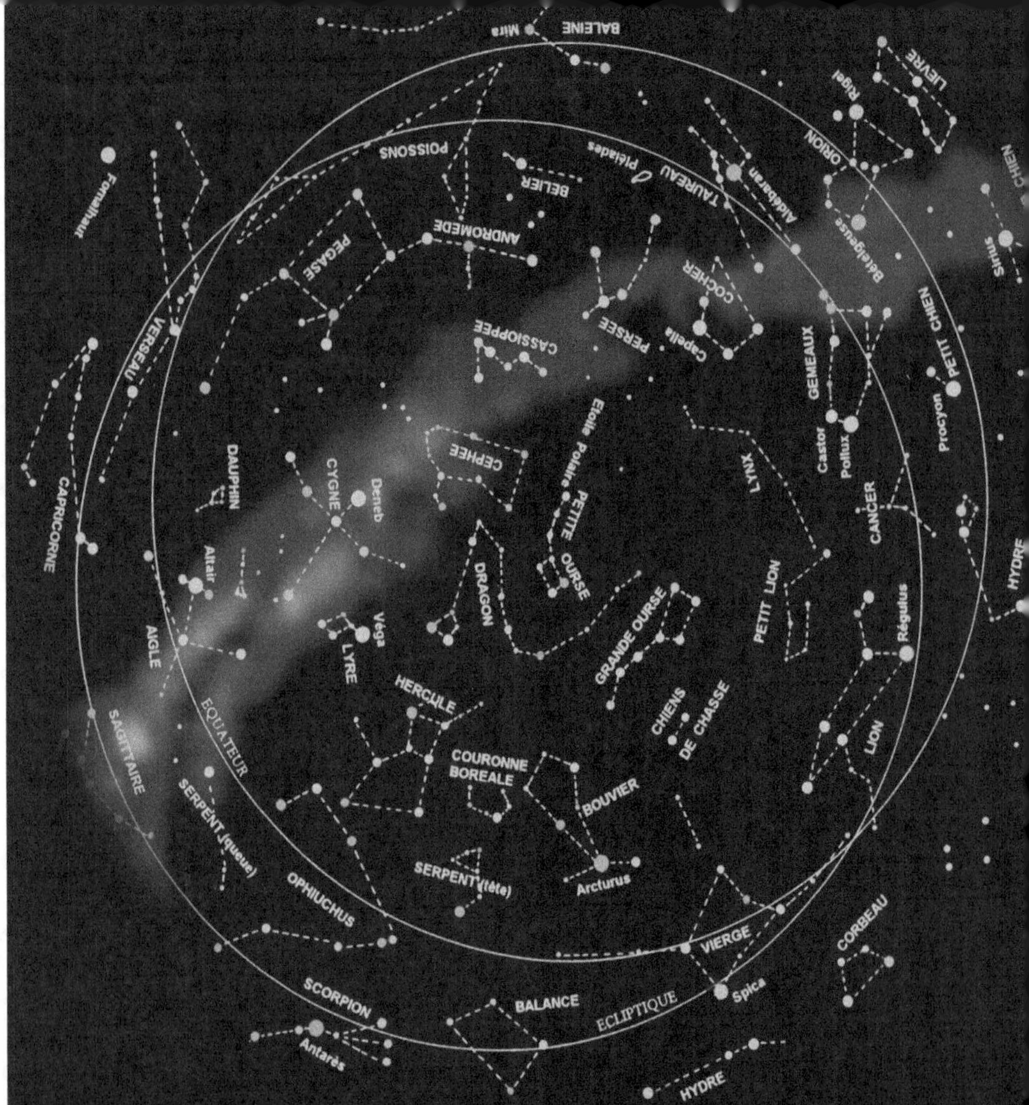

Super Networked Business of the

Third Richest Man on Earth

Chapter 13

Super Networked Business of the Third Richest Man on Earth

> ***"Wealth is the product of man's capacity to think."***

Ayn Rand (1905-1982)

> ***"A creative man is motivated by the desire to achieve, not by the desire to beat others."***

Ayn Rand (1905 - 1982)

He grew up facing grinding poverty and impoverishment in rural areas of Spain. If someone had told Amancio Ortega that one day Fortune magazine from the far-off land of America would herald him as the third-richest man on earth, or that his business network will directly or indirectly affect the lives of more than one million people on earth he would had laughed with disbelief. The son of a housemaid, he lived the first decades of his life in quarters abutting railway tracks and would at one time have considered himself lucky to join his father in a railway job. Instead, he embarked on a journey that led him to build a global network of fashion businesses and become the third-richest man on earth, surpassing the much acclaimed American investment guru Warren Buffet.

5-STAR Business Networks can create wealth beyond dreams

Heralding his reign as the third-richest man on earth, Fortune magazine (January 8, 2013) declared:

> *And yet, a world away from the glitz of Paris, Milan, and New York, Ortega has built a fashion empire that reaches into more than 80 countries. Beginning 40 years ago, Ortega ripped up the business model that had been refined over decades by Europe's fashion houses and replaced it with one of the most brutally fast turnaround schedules the industry had ever attempted. Decades later Zara is the world's biggest fashion retailer.*
>
> *Ortega built his empire on two basic rules: Give customers what they want, and get it to them faster than anyone else. The twin organizing principles have made the company (and Ortega) into an unlikely iconoclast, more of an optimal supply chain than a traditional retailer. They are also the secret to Inditex's astonishing success* [1]

It all started in 1949, when Amancio Ortega Gaona began his career in the textile industry for shirtmarker La Coruna. Even as an errand boy, Ortega learned how costs added up through the apparel value chain. Working out of his sister's home, he began assembling his own designs. By 1963, Ortega left his job and founded his own company called Confecciones Gao with just 5,000 pesetas (equal to $25).

Wealth is the product of man's capacity to think

Ortega's capacity to think up a new business model for the fashion industry led him to reproduce popular fashions using less expensive materials in order to sell high-demand clothing items at lower prices. The business ran smoothly, manufacturing lingerie and nightwear, until 1975 when a wholesaler cancelled a large order in which Ortega had all his capital tied up. Besides going bankrupt, Ortega had no other choice than to open his own store. He called it Zara because his preferred name - Zorba - was already taken.

It helped that until the end of the 1970s the Spanish textile industry was highly protected, oriented mainly towards domestic markets and with an extremely restricted export capacity, regardless of state aid and the competitiveness of its products from low wages. From these early 'incidents' in his career, he discovered that to be successful "you need to have five fingers touching the factory and five touching the customer." (Erikkson, C., & Jonsson, S. (2011)[2]. Inditex: A company analysis with focus on growth)."

Of those early days Fortune (January, 2013) says:

Ortega had the ideal environment: Galicia. With few job opportunities, thousands of men worked at sea, leaving their women to struggle alone back home. "The women would do anything for a little money, and they were really good at sewing," says Blanco, who co-wrote a book called Amancio Ortega: From Zero to Zara. Ortega began organizing thousands of women into sewing cooperatives. He oversaw a thriving production of quilted bathrobes for his first company, GOA. Mercedes López was 14 when she went to work for Ortega and says most women were thrilled to be hired. "The conditions were really pretty good," says López, now 52, who is the textile union representative at Inditex. "We knew Amancio well. He was very close to the workers." It was a family business: Ortega ran design, his brother Antonio headed the commercial side, and his sister Josefa was the bookkeeper. The company trucked in textiles from Barcelona, cutting out the middlemen." [1]

Network of sewing co-operatives became the bedrock of Zara's success as it expanded its sales footprint

Zara became well established and quickly expanded to larger Spanish cities. Zara represented in Spain an essential departure from the prior notion of fashion. In a way, Spain was going through the process of democratisation in the post-Franco era and saw democratisation of fashion through Zara since, from these stores, chic couture with innovative designs was accessible to the majority of pockets. Until then, design-intensive textile products did not stem from mass industrial production and were affordable only to select few customers with high buying power.

In 1988, Zara introduced its first store abroad in Oporto, Portugal. Shortly thereafter, stores were opened in New York and Paris. Since then, through a relentless globalisation drive to build a 5-STAR Business Network, Zara products are offered in 77 countries with 1,518 stores, comprising about 64 percent of Inditex's total business.

Without a doubt, Zara is the leading brand in the Inditex Group. Zara strategy is based on the idea of sharing the same clothing culture all over the world with a slight adjustment to cater to the differences between countries.

New brands to expand the reach of the network

Kiddy's Class was established in 2002 as an independent fashion outlet exclusively aimed towards kidswear. It provides the most modern designs and quality apparel at reasonable prices. The brand, with stores primarily in Spain and Portugal, grew out of the home market in 2006 by launching newer stores in Italy, France, and Greece. Currently, however, Kiddy's Class is not an independent brand; it is grouped with **Zara Kids**.

Every year the company has augmented its presence,

reach and profitability through smart creation and application of a 5-STAR Business Network. In 2007, Zara launched a collection of maternity wear. Presently, the Zara for mum line is available in all countries around the world where the brand is present, offering a wide range of garments.

Another brand - **Pull & Bear** - was initially created in 1991 and was focused on providing basic men's apparel only. Within the next two decades, the product range extended to not only women's fashion but also to perfumes, accessories, and home decoration. By now, the brand consists of 682 stores in 46 countries across the world, amounting to $857 million dollars, which embodies 6.8 percent of Inditex's total net sales.

Brands created their own pull with their target customer segments via customer intimacy strategies

In 2007, Pull & Bear created a blog on its website. The purpose of this blog was to act as a meeting point for the global youth culture where customers could update on the newest trends in fashion, music, and visual arts. In addition with the association of the youth culture, Pull & Bear also participated in environmental and charity projects, many of which appeal to the youth. For instance, a forest project was established to avoid the cutting of trees and as an outcome, the brand discontinued the publication of its catalogues on paper. In addition, the brand contributed to the reforestation of the Sierra Gorda biosphere reserve in Mexico. Lastly, Pull & Bear built an interactive micro-site where customers can discover environmental concerns through the assorted games and advice.

About one year later, Pull & Bear generated another project. Combining ideas of art, sport, social and environmental activities, the brand portrayed a major social project called the Pull and Bear world project. Overall, understanding music and chasing dreams are the common denominators of many young people is the main reason why the company's relationship with its customers is significantly strong.

Careful market segmentation results in significant market coverage with allied brands

The third brand Inditex became interested in was **Massimo Dutti**. The first store was opened in 1985, but it was not until 1995 when Inditex obtained 100 percent ownership of the brand. Massimo Dutti was primarily a line for men's clothing and accessories, but when the company took over, it expanded into women's clothing and accessories as well. With 530 stores, this brand operates in 50 countries across Europe, Asia, America, and Africa. The brand's annual net income is around $897 million dollars and makes up 7.2 percent of Inditex's total.

The design appeals to men and women from the urban backgrounds, while also catering to a casual style. Generally, the Massimo Dutti look is the concept of a universal design that is independent and cosmopolitan. The fashion line symbolizes comfort and adaptability for modern individuals - appropriate for any circumstance.

In typical Inditex style, the brand expanded into a collection for children called Dutti boys, in 2003. Unsurprisingly, the collection was later broadened to clothing for girls and renamed Dutti Boys & Girls.

One year later, Massimo Dutti created a unique offering - a personal tailoring collection. This let customers design their own garments deciding which fabric and colors they wanted.

Each of the brands builds bold links with its key constituency

Started in 1998, **Bershka** is another Inditex brand. This label originally began selling daring clothes for women but branched out and offered clothing for men as well. The overall collection focused on the youth culture, between ages 13-23. Bershka is composed of 720 stores in 50 countries. Net sales represent 10 percent of the Inditex group, making it the second-largest brand in the Inditex Group.

Bershka's concept is slightly divergent from the other brands not only in the layout which portrays a more modern store, but also in providing an overall socio-cultural shopping experience. An urban, fun environment with a unique image is what the brand aimed for. The stores are roomy and act as a hub for fashion, music and art where young customers can gather. The stores allow customers to read magazines, listen to music and study art.

Desiring further change, the brand reformed its image and renamed it Magic in 2009. All stores altered their look based on a digital world with a dark color scheme, pale lighting, and house-dance music, while still offering latest fashion trends and bold clothes.

At times same demographic is served by two or more brands - catering to different tastes

Powerful brand **Stradivarius** was launched in 1999. Following the Zara model, the line concentrated on clothes for young women between the ages of 15 and 25, but with an added twist. The fashion concept focused on the newest trends in international fashion and its design, fabric, and colors are illustrated by an informal and imaginative style. Stradivarius is active in 43 countries with 593 stores and comprises 6.2 percent of Inditex's total. In 2007, the brand revamped its image by creating a commercial message: Take me or leave me.

In 2001, **Oysho** was opened and provided the hottest fashion trends in women's lingerie. The collection included underwear, sleepwear, swimwear, accessories, and cosmetics, among others. With 432 stores in 25 countries around the world, Oysho represents 2.4 percent of Inditex's overall size. By 2007, a line for babies and girls was launched and named My First Oysho. The collection of underwear is directed at children between the ages of three months and nine years.

Additionally, another line was established called, Homewear, consisting of clothing that could be worn at home.

Growth was fuelled by going beyond apparel

Continuing with the idea of home goods, a brand of typical and modern household products was launched in 2003. Zara Home maintained the same concept as the other fashion lines of the company. This was also the first time Inditex ventured out into the world of e-commerce. Zara Home has 284 stores in 27 countries making up of 2.3 percent of the total for Inditex.

In co-operation with the original Zara, the brand also participated in two other crucial projects, constructing an eco-efficient store and introducing the application for smartphones.

Last but not least is Uterque, the newest brand established in 2008. The brand concentrates on sophisticated fashion complements and accessories created by recyclable material. The brand offers handbags, footwear, jewelry, and leather garments. Forming a miniscule 0.5 percent of Inditex's revenue, the brand runs 80 stores in 16 countries.

Overall, Industria de Diseno Textil (Inditex) is the holding company of the eight different concepts. Since 2005, Inditex is a world leader in the fashion industry.

The company claims, "it is possible to both grow and maintain profitability levels by making the most of the competitive advantages of our business model as the group becomes larger and more complex." Additionally, Inditex declares that its fashion concept is linked closely to a well-developed management structure, which makes possible the rapid growth of Inditex (Erikkson and Jonsson) [2].

All the brands are serviced by common portfolio of 5-STAR capabilities

In 2003, a logistics centre, called *Plataforma Europa*, was constructed in Zaragoza, Spain, to enhance the flexibility of Zara's logistics system and strengthen its distribution capacity. The key reason was to continu expansion of Zara into global markets that needed even more evolved logistics capability. The creation of this centre amplified the handling of international cargo and permitted a distribution capacity of 80,000 garments an hour.

To enhance customer intimacy, Zara acquired the capability to create a shop window via social networks displaying the most up-to-date details of new trends. Its application for the iPhone and other smartphones enhance this capability.

Besides launching new brands, during this high growth phase, Inditex was also listed on the Spanish stock exchange, permitting shareholders to invest in the company.

Fire-Aim-Ready (FAR) Innovation keeps the stock fresh and crisp

Inditex is a leading retail innovator and the company stresses that innovation is completely fundamental because Inditex's business model is based on innovation and flexibility. Its growth as a group is based solely on its capability to innovate and that's exactly how it broke numerous moulds in the fashion industry.

Ortega truly believes that creative imagination is the root of innovation. As mentionned earlier, one quote he lives by is: "When all think alike, then no one is thinking." (Walter Lippman)

Starting off, Ortega realised that it is not only the rich that appreciate expensive and well designed clothes. As a result, the concept of creating high quality, trendy clothes affordable to all became Ortega's target. Therefore, this original notion fits well with Inditex's slogan, "We seek to democratise fashion, in contrast of the idea of fashion as privilege" (Small Business Database) [3.]

More than anything else, the success of the company is built on its rapid design cycle

The group's Zara chain is built on the idea of turning over the product range every few weeks. In an industry where most competitors outsource manufacturing to low cost countries, Inditex is intimately involved in all stages of the production process: from design, manufacturing and logistics, to sales in its own stores. The entire process takes two weeks; on average, Inditex produces 80 percent of its merchandise in Europe. The result is a higher labor cost of about 15 percent over its low-cost sourcing rivals. Ortega nevertheless generated competitive advantage by maintaining lower inventory and minimising shipping costs by keeping production close to the markets (Small Business Database)[3.]

The company modernised the textile sector with the idea of bringing out a number of diverse collections each season instead of the standard two collections each year. The Zara model came to be acknowledged and noted in business schools as a successful alternative to the existing model that called for complex supplier networks taking up to nine months from the design to the stores. "It's really quite exciting to see a new model like this - one that actually breaks all the rules of the apparel industry," stated Jeffrey Ballinger (Small Business Database)[3].

Just like Dell and Amazon, Inditex too stood a traditional industry model on its head

Compared to the traditional model for innovation, Inditex's model is quite the opposite. Instead of the cycle being design, source, store, then sell/customer, Inditex reverses the cycle - going customer feedback, design, sourcing, produce and sell.

In another break from the industry tradition, Inditex has a policy of zero external advertising, and instead relies heavily on word-of-mouth. Money is invested into new store openings, sustainability, and fresh store design; the company focuses on marketing inside the store (GO Boldly)[4]. In 2012, Inditex was announced as one of the most innovation leaders across 25 sectors every year since 2001.

Among the eight brands within the Inditex Group, not only is Zara the top innovative retail group within Inditex, but also in the world. Zara worked hard to get to the front and concentrates on vital fundamentals. Zara acquires design ideas from diverse sources, including books, magazines, Twitter and blogs, as well as fashion shows and forums - the brand searches for different visions all over the globe.

The feedback from the stores is a critical component; it is a crucial area the design team focuses on as consumer demand is a paramount aspect of business. All and all, the Zara design team of 350 people is responsible for creating and delivering 18,000 new product designs a year, which boils down to more than 70 distinct pieces a day (Sloane)[5]!

The speed of innovation is aided by technology

When it comes to use of technology to aid industrial production, Zara is a natural leader. Moving quickly from design to market is what separates Zara from its competitors. While the production timeline for its competitors can take as long as five months, turnaround time for Zara stores is usually only three weeks. Zara moves much faster and creates waves in which others follow. The brand puts products into 5,500 stores from Brisbane to Lisbon at just about the same time. It makes a small number of adaptations for local markets. Zara products are designed with a global mindset and sell on all five continents, giving enormous economies of scale.

Each Zara store gets replenished twice a week, based on real-time inventory data collected at each store and sent to computers at the company's headquarters. Inventories are kept low whereas fresh designs pour into stores almost constantly. Technology plays an important part in the group's ability to innovate.[5]

The innovation strategy Zara depends on is rapid prototyping and modeling whereby designs and ideas are tested quickly. Zara's headquarters in Arteixo La Coruna, Spain, contains 25 shop windows of full size displays with controlled lighting. The designers can quickly assess the appearance of retail shop windows by simulating different times of the day and weather conditions. (Sloane)[5].

The culture of Inditex is steeped in innovation

Innovation at Zara does not apply exclusively to products. One lesson Zara learned is to innovate in all aspects of the business. It has used creative methodology to restructure and optimise manufacturing, supply chain management, data tracking, inventory management, store layout, and staff operations (Sloane)[5].

Fast feedback is a major enabling factor for innovation for Zara. The store staff uses wireless communications to send information every single day back to Spanish headquarters about sales and inventories. Relaying fast information about what is selling and what is not assists the group to react speedily to public behavior. Furthermore, reports facilitate faster supply of the most popular trends to where they are in demand. Strikingly, even the top selling fashions are withdrawn after four weeks; Zara intentionally does this in an act of creative destruction. As a result, customers must visit the stores frequently to guarantee that they do not miss out on the latest items (Sloane)[5].

Zara's organisation structure aids innovation. Inditex does not foster a hierarchical system of management. With a relatively flat structure working closely between the various parties akin to Figure 8.4 on page 160, Zara is able to rapidly design, procure, produce, transport and sell product even before competitors send the designs to the production floor. With this arrangement, designers are encouraged and motivated to make fast decisions. All and all, Zara's successful growth is not purely a result of low cost or of standardisation but rather of innovation (Sloane)[5].

Innovations like these have stimulated Inditex to grow from $2.43 billion in 2001 to $13.6 billion in 2007 and have sparked enormous global expansion, opening hundreds of new stores around the world. Inditex's skill to endlessly launch new products and then improve them quickly using customer feedback has paid dividends.

$eed-to-$tore (t) Efficiency allows Ortega to 'democratize' high fashion

While embracing innovative management practices to improve performance, Inditex also works on becoming more efficient and sustainable. For example, compared to its key competitor, H&M, Inditex's operating profit margin is 21.6 percent while H&M's is 13.1 percent. Inditex is more efficient in producing a larger profit per euro of sales. Inditex's higher operating income is an outcome of keeping the costs of goods sold and operating expenses much lower than (Kihlen)[6].

Inditex's reduced costs are possible because of in-house production, lower advertising expenses and keeping a cost-effective number of employees per store. H&M has a higher number of employees per store (29.7 percent when compared to Inditex's 20.8 percent). This is partly responsible for H&M's higher cost of goods sold.

Inditex's unique business model creates efficiency

Fewer and larger factories allow Inditex to benefit from economies of scale. Additionally, to cope with Inditex's higher replenishment frequency and wide store selections, it has invested in highly flexible machinery and extremely experienced workforce in order to supply apparel in a fast and efficient manner.

Due to the location of its factories and sewing co-operatives in Europe, production costs are fairly high for Inditex's supply chain in contrast to their competitors (Katigbak, Labog, Leung, Li, Ortiz, and Papilla)[7]. However, this allows Inditex to build and operate a centralised distribution center from Europe. Coupled with its customer intimacy, rapid product deployment in store and resulting high margin, the overall formula is quite successful.

Inditex is not looking to shave every last cent out of its production cost. That would mean outsourcing to far off lands, long supply chains, fewer number of fashion cycles per year, much longer fashion cycles, deeper discounting towards the end of each cycle etc - all ills that plague its competitors in the garment trade.

On the contrary, Inditex has stuck to its formula of very short fashion cycles, rapid product development and deployment, fresh crisp stock in store every few weeks, almost no repetition of designs and keeping the customers excited and in suspense about what new items they will find in the store. This unpredictability is itself the source of endless discussions among its loyal fan club and has buyers on tenterhooks, keeping the cycle running over and over again.

Inditex invested in technology, systems and intellectual capital to maximise the t Efficiency within this formula of operation creating a unique model of efficiency.

As the company expands it footprint, the networks continue to evolve to create further $eed-to-$tore (t) efficiencies

Inditex is confronted with a growing pressure from its global expansion. Limits to growth in its traditional markets means it can grow at its usual pace only by bringing its unique model to other geographies. However, with the expansion of Inditex brands in Asia and North America, come growing pains. With the manufacturing still centered around Europe, lead times to the stores in far off geographies can become too long. Moreover, the competitive advantage of Inditex rests on very intimate understanding of trends and currents of its customers' tastes and desires. With growing distance and size, this intimacy could diminish.

Many analysts have expressed concern that transportation costs will rise given that the product needs to travel longer distances. Zara has an exceptionally effective and efficient model in its home European market. It will need to make modifications in its supply chain consecutively to expand to other markets around the world (Katigbak, Labog, Leung, Li, Ortiz, and Papilla)[7].

Another concern expressed by the analysts related to a single distribution center is the possibility of diseconomies of scale. As of now, Inditex is tremendously rapid and efficient with its distribution center. No inventory stays in the center for more than three days. However, there is a limit to cross docking in a complex operation - at a certain volume the resulting confusion can easily blow out the time in warehouse. The overload volume may surpass the working capacity of the capital equipment within the distribution center and lead to chaos. Additionally, analysts have also expressed concerned about choke points in a single distribution centre and highly concentrated production. As there is only one main distribution center, any disruption would be highly risky to Inditex's business. Therefore, as per these commentators an additional distribution center would be valuable not only because it can help cope with the extra demand brought about by expansion, but also because a single DC to handle all of distribution is risky. (Katigbak, Labog, Leung, Li, Ortiz, and Papilla)[7].

Of course we cannot predict what will happen in the future. From all indications the company seems to be adapting to its growing global footprint quite admirably. Despite warnings from analysts, some of which appear to show self-interest, Inditex has managed to adapt to its global expansion reasonably well.

How?

We will examine this in more detail later in this chapter when we look at the ROM techniques used by the company. The same strategy it used for modularised sewing co-operatives in Europe has been adapted on a much grander scale at a global level.

The business network is constantly under rigorous review

As the company expands into North American markets, distance from the manufacturing and distribution hubs in Europe puts additional stress on the business model. What could be designed, manufactured and put into the stores in three weeks suddenly takes 8-10 weeks. There are cheaper production and distribution locations in Latin America, closer to the North American markets.

Analysts and magazines openly question the wisdom of the traditional Zara model for the globalised company. Manufacturing and distribution out of Europe seems to make the costs excessive while reducing the supply chain flexibility and capability to respond to the market trends. Another distribution center would reduce the economies of scale, but if the market in the other country were large enough, the decreased transportation costs would prevail over the increased warehousing costs. One country Inditex could establish another distribution center is Mexico. Labour costs would be low and the location would be perfect for delivery to North American markets (Katigbak, Labog, Leung, Li, Ortiz, and Papilla) [7].

While the analysts continue to comment and suggest network optimisation moves, Inditex has used its tried and tested model of staying close to customers and making customers centre of all strategic decisions. If that means more diversified production or distribution, Inditex may not hesitate to make those moves. Meanwhile, everyone watches the newer network configurations emerging out of the company headquarters with great anticipation.

Manufacturing costs in traditional locations are now comparatively higher than low-cost countries

It is well acknowledged that the manufacturing costs in Europe, even using the traditional sewing co-operatives and job-shopping workshops that Inditex pioneered during the 1970s, are now far more than encountered in factories in China, Vietnam, Bangladesh and Pakistan.

For Inditex, the key competitive advantage remains in being able to adjust product attributes quickly to keep its fashion line fresh and crisp.

Inditex still creates low overall costs by efficiencies

With up to 18,000 new styles being made and introduced over a year, every style, on average, earns total revenue of less than $1 million each. This figure is much lower than other retailers or brands and appears, prima facie, not cost-efficient in terms of design and product development costs. However, higher cost of product development is more than sufficiently rewarded by higher realised margins (Plantilla Simple)[8]. Therein lays the elegance of the Inditex business model and its 5-STAR network. The competitive advantage is not in the lowest costs because others could come in and outsource to the same factories in the Far East. The highest overall margin is the source of its competitive advantage, gained through a tight t Efficiency.

Transaction Optimisation Profitability (TOP) is achieved through customer intimacy

There is no doubt whatsoever that Inditex is a customer-driven company, from its very core to outermost periphery of its 5-STAR network. It has managed to focus the entire network solely on the customer, hence the importance placed on its customer facing units. Inditex believes that it cannot let its stores down. The company replenishes store stock twice a week and rolls out new collections every few weeks. As a result, it is necessary it has an efficient logistical system (Plantilla Simple)[8].

While concentrating on maximising efficiency, Inditex takes profit optimisation extremely seriously. Maximising revenues and minimising costs to achieve the best possible overall profit is every business's goal. However, Inditex manages to achieve it annually.

The company's profitability analysis, and the fact Ortega is now the third richest man on earth, surpassing Warren Buffett, tells us that the results of the company are winning those accolades.

The strategy Inditex utilises to earn the largest profit for each transaction is an integrated cost leadership/differentiation strategy. It offers current and desirable fashion goods at relatively low prices. In order to execute this strategy effectively, involvement of sophisticated designers and effective means of managing costs is necessary. Inditex embraces these capabilities constantly (Hitt, Ireland, and Hoskisson)[9].

Generally, an integrated cost leadership/differentiation strategy can provide a company a competitive advantage. First, minimising costs enables products to be offered to the market at an incredibly low price. And secondly, the differentiation enables it to present unique value for which the consumer is ready to pay equivalent higher price. When the level of differentiation is crystal clear, the price premium can be surprisingly high. To enable product differentiation, it is also essential to effectively segment the marketplace, target the most attractive segment and position the company and its product appropriately.

In fashion markets, the consumer searches for a product with social significance; but at the same time a product that is functional and above all one which has a distinctive style. This distinctive style is one of the criteria that allow the consumer to recognise and distinguish one offer from another (Dopico and Porral)[10].

Inditex knows its customers well and crafts careful offerings

Today's consumers are spending less on clothing - choosing to use their disposable income on healthcare, electronics, education, and travel and leisure. Furthermore, they have ability to find pricing and product information on the fly using internet and smart-phones, and can now have a much wider and global choice from a variety of inexpensive channels.

To capture today's indefinable consumer, it is more central than ever to find a breakthrough in the clothing market. Therefore, Inditex has found differences that matter to customers and differentiated itself from its competitors by executing vital activities in its supply chain differently. It is this that sets barriers for competitors because they will find it difficult to imitate Inditex's positioning (Pearson)[11].

Aside from determining the demographics of the customer, the company also targets customers based on their sense of fashion and style. The ethnicity of the brand and its target market is merged by Zara in its product offering, which corresponds with an array of tastes and settings.

The stock is kept fresh and crisp - and rarely repeated. This maximises revenue due to less discounting

Furthermore, the company aims to attract customers who are attracted to high fashion and the latest trends but cannot afford clothes from the high-end boutiques. In order to target the market, Zara's strategy presents its outlets in high profile locations and supplies customers with a turnover time of 4-5 weeks for its new collections made accessible at a fraction of the couture cost. This, along with the brand persona, the collection of the clothes and accessories and the marketing operations pulls the target consumers to the Inditex stores. In this way, Inditex achieves differentiation that is sustainable, built upon the company's ability to design, manufacture, and stocks the stores with the newest fashion styles in a short amount of time of two weeks and at a significantly lower price compared to its competitors. Inditex proves that differentiation can yield higher sales and less margin risk (Jannoun)[12].

To optimise profitability, costs are tightly controlled

None of its competitors come close to maximising net working capital as Inditex does. The key to Inditex's success in managing its working capital efficiently is its quick response system, which lessens cycle times and permits the company to closely match supply and demand, keeping net working capital requirements at a minimum. Since inventory is one of the main components of current assets, and forms a critical factor in the net working capital formula, it is evident that Inditex has the ability to minimise inventory costs, allowing it to attain far greater operating efficiency than its competitors (Inditex) [13].

As mentioned earlier even though the labour cost in Europe is higher, the efficiency of this system allows Inditex to keep costs down by spending less on transportation and keeping inventories low. A sophisticated optimisation model is utilised to maximise its global sales by varying the inventory of each piece in the warehouse and each store, and the assortment decisions as inputs to the optimisation model, and replenishment quantities as its main decision variables (Caro and Gallien)[14].

Inditex does not experience loss when entering a new market, as the company does not have extraordinary advertising expenses to create brand recognition. In contrast with other clothing retailers, who spent 3 to 4 percent of sales on advertising, Zara spent simply 0.3 percent. The little bit that the brand spends goes to support its identity as a low-cost but high fashion clothing retailer (Pearson) [11].

Sensible pricing strategy maximises profitability

Additionally, Inditex's pricing strategy is to set price equal to cost plus a target margin. "Inditex prices are based on comparables within the target market, subject to covering costs plus a target margin." Inditex has a low price strategy to take advantage of its lower cost structure when compared to their high end competitors with similar product range. The quality of the products is unique and they can cut costs, enabling it to offer a lower price (Shehzade) [15].

Its pricing power bestows on Inditex a cushion partly responsible for its success. Careful market segmentation and pricing to suit each market segment, utilising its various brands, enables Inditex to offer products and prices suitable for both wealthy and non-wealthy consumers.

Being a data driven company, Inditex is able to track the profitability to lowest level of transaction possible. As a result Inditex achieves 15.7 percent on its return on assets with help from the responsive supply chain. Additionally, the company achieves 24.8 percent on its return on capital and 30.4 percent on its return of equity (Bloomberg Business Week)[16]. These spectacular results show the power of its 5-STAR Business Network in action.

Advanced Product Phasing (APP) allows Inditex to launch a barrage of fashion cycles - one after another

We have already seen how successful Inditex is in product design through FAR innovation strategy. Inditex starts this process by receiving feedback from the stores. Reviewing sales reports and qualitative comments, cross functional teams focus on the responses that will be the seeds for the next cycle of innovation.

In addition, market research is a key component for the company to find out who the early adopters are. Collecting data from university campuses, nightclubs, and all the other places frequented by its target segments; the information is then conveyed daily to headquarters (Brand Watch: Zara)[17].

The trend information is collected anywhere and everywhere. An example quoted by Fortune magazine showed how Ortega operates:

> *The motorbike roared up to the traffic light in La Coruña in northern Spain and stopped alongside a black Town Car. From inside, the passenger glanced out his window and saw the young biker leaning over the handlebars, jean jacket decorated with appliquéd patches, a throwback to the 1970s. The man in the car, decades older than the biker, zoomed in on the jacket. The old man grabbed his cellphone and, as the story goes, called an aide in his office. His eyes still fixed on the biker, the man described the jacket's stitch-*

ing, its shape and color, and signed off with a single instruction: "¡Hácedla!" Make it.

The light turned green, the biker pulled away; unbeknown to him, he and his jacket had just played a walk-on role in one of the greatest retail stories of our time [1].

Inditex observes the trends and buying behaviors in great detail. For instance, if a design does not sell well within a week, it is automatically removed from shops, advance orders are cancelled and consequently, a new design is engaged (Management for Marketers) [18].

Core of Inditex is its legendary design capability

Next, Inditex concentrates on quick decisions and designing the product. The feedback from around the world is collated and analysed at its global headquarters.

At headquarters, the design centre includes three big halls, one for each product line - men, women, and children. Inditex does not save money by providing the same labour for all product lines. Each line has its own devoted staff members, from design to procurement (Erikkson and Jonsson)[2].

With the three design teams, working close to one another, the speed and quality of the design process improves. All the designers can easily check on each other's sketches. In the corner of each hall, there is a minor prototype store set up where the designers can put up creations and receive criticism from the other designers and market specialists. The market specialists note the new designs' style, fabric, color, and can propose retail prices. The procurement and production planners estimate production costs and available capacity and can, if desired, begin the production of a new garment in just a few hours (Erikkson and Jonsson)[2].

Generally, the designers at Inditex produce about 40,000 new designs annually; only a third to half of them are chosen for production. Every year, 12 to 14 collections are introduced rather than the traditional fall and spring collections (Erikkson and Jonsson)[2]. Therefore, the product life cycle is deliberately curtailed in a manner that keeps the consumers on tenterhooks, forcing buying decisions instead of procrastination and building excitement about the fresh and crisp product line.

Designers are closely aligned with manufacturing

The majority of Inditex's manufacturing is done at its own factories, particularly for the high fashion items. Only the basic clothing is outsourced because there is no issue with seasonal change. When developing a product, Inditex uses a process that is driven by the demand "pull" instead of traditional "push" products to the market based on demand forecast.

Furthermore, Inditex directly controls fabric supply, marking, and cutting. For the sewing co-operatives and external workshops, mainly in Spain and Portugal, Inditex supplies fabric and all the required components for making pieces, unlike H&M, which outsources all production and does not directly work with material suppliers. This benefits Inditex by eliminating the need for information to travel through scattered channels.

As a result, the designers work closely with the manufacturing and are closely aligned with the marketing and sales. This reduces delay, minimises bureaucracy, permits faster decision making and decreases the impact of supply chain mismatch like overproduction. Lastly, Inditex accomplishes speed and flexibility by purchasing half of its fabric plain and undyed. The outcome helps the business lower the risk of incorrect forecasting of seasonal colours and react rapidly to colour demands (Zara Production) [19].

Network of smaller workshops play a crucial role in its success

Labour intensive steps, like sewing and finishing of the products, are positioned on a network of around 450 small workshops located in Galicia, Spain and northern Portugal. Nearly all of these workshops utilise around 20-30 staff each, and are specialised by production process or product type. The finished products are then sent back to Inditex where they are quality controlled, ironed, labeled and packaged. Inditex is largest and most significant customer of these workshops. Thus they normally keep long-term relationships while Inditex offers them technological and financial support (Erikkson and Jonsson)[2].

All and all, the product facilities hardly ever run at maximum capacity. Instead, they manage simply a single shift, so they can improve production and meet both expected and unexpected demand increases when considered necessary (Erikkson and Jonsson)[2]. Inditex stresses that designing and responding quickly is more central than sourcing quickly.

Even its distribution strategy is aligned to Advanced Product Phasing (APP)

The Zara stores receive shipments of products twice a week, which they can put on display instantaneously as they are already ironed and some are shipped on hangers. As a result to these continuous shipments of new items the Zara stores can hold fewer inventories than its competitors.

The items shipped to the stores are mainly based on the store manager's judgments. It is the store manager's responsibility to keep track of which items need to be restocked based on popularity and missing sizes. The store managers get information once a week with only the details of the items in the central warehouse while the quantities are not specified. After that, the managers send their requested shipment qualities to the warehouse. These numbers are tailored by the warehouse employees to allow for the total outstanding quantities at the warehouse and the wanted quantities from each store (Erikkson and Jonsson)[2].

Its control of the entire business network allows it to create a unique product life cycle profile

Overall, Inditex runs a highly integrated vertical supply chain all the way, from raw materials supply to its retail stores. The company controls most of the steps on the supply chain from designing, to producing, to distributing. Inditex's business system is unlike competitors and is known for manufacturing its most fashion-sensitive products internally. Inditex does not make "classics" or clothes that will forever be in style. Instead, the company aims for its clothes to have fairly short life spans, both within stores and in customers' closets (Mukherjee, Sujit, Janmeja, Mohata, Rani, and Shinde)[20-21].

Inditex's product life cycle is exceptionally divergent from the norm. In a typical product life cycle curve, sales usually decrease as the product moves over time. However, with Inditex being in a high fashion industry and its product offering being the newest trends and designs, the maximum life cycle is 5-6 weeks for each piece (Mukherjee, Sujit, Janmeja, Mohata, Rani, and Shinde)[20-21].

Inditex's unique product life cycle profile is one of the bedrocks of its success

The normal time frame for a product consists of one day for the actual style idea, another day to obtain fabric from stock, an additional three days for design, an extra day for style approval, three days for the prototype development, ten days for garment product, and lastly, five days for shipment (Brand Watch: Zara)[17].

Shortening the product life cycle means larger success in meeting consumer preferences. No design stays on the shop floor for more than four weeks, which motivates Inditex customers to make repeat visits. An average high-street store in Spain anticipates consumers to visit three times a year. For Inditex, it's 17 times (Management for Marketers)[18]. Overall, Inditex is known for its rapid, yet winning product life cycle management, highlighting its APP capability.

Results-focused Outsourcing and Modularisation (ROM) allows Inditex to grow rapidly and yet retain flexibility

There are various definitions of supply chain management but one is an effort to work together with all the network's partners to support the flow of material up to the customer. To have the capability of creating and placing new designs or styles in the stores in extremely short periods, equal dedication from the outsourcing and partner companies is necessary. The function of supply chain management within the fashion industry is to be able to distribute the fashionable styles at the time insisted and in the right place (Choudhury and Holmgren)[22].

As the company globalises, it needs to outsource more

Globalisation has become impossible to avoid over the past decade for any business. For fashion retailers it provides sources of inspiration for new design, as well as growth markets and economies of scale.

With globalisation comes pressure for supplier selection, development, and management, as retailers head towards global outsourcing. Given this competitive situation and the promise of cost efficiency from Low Cost Country Sourcing (LCCS), sourcing from the Far East is very popular with European or American fashion firms (Choudhury and Holmgren)[22].

The outsourcing method being largely used by European and US apparel retailers is global outsourcing to achieve higher profits by cost reduction. This is done by rigorous vendor selection and management via closely monitored contractual relationships.

In order to combine the benefit from outsourcing globally, a balance is essential between the global and local sourcing for the greatest profitability; there are benefits to either approach. Low cost country sourcing offers tantalizing cost reduction while, on the other hand, local sourcing offers differentiation capability and agility.

With its global expansion, Inditex needs to outsource even more in order to create and harness a 5-STAR network effectively.

Inditex brings its own unique touch even to outsourcing

However, Inditex utilises a unique strategy for outsourcing - its blend of the global and local sourcing. Half of the products Zara sells today are produced in-house and the other half of the products are split equally between other countries in Europe and Asia, (Goransson, Jonsson, and Perrson)[23]. Inditex outsources less production than its competitors. Inditex outsources some of its regular products like cotton T-shirts, which are price-sensitive to Asia as these products have longer shelf lives. Majority of the Inditex's supplier factories are in Spain and Portugal and these factories focus on fashion-customised sewing.

The closeness of these factories, specialised in the manufacture of fashionable garments, to the distribution facility saves time and transportation costs. This allows Inditex to respond rapidly to fashion trends, therefore, decreasing its product risk (Inditex: A strenuous supply chain). Inditex has built a unique business-model and today Inditex is a vertically integrated retailer. The business manages every step of the value chain and is continually researching, making it an ongoing development (Goransson, Jonsson, and Perrson)[23].

While it does not own the workshops, Indetex controls the entire business network effectively

Today, a significant proportion of fashion companies have their products made in China or other such low cost countries. Although this strategy is apparently cost efficient, controlling a long supply chain is notoriously difficult. By the time a boat has sailed halfway round the world, hemlines may have risen an inch and its cargo will be as admired as geriatric haddock. Instead of gambling on tomorrow's latest look, Zara waits to see what customers are buying and then quickly makes more (Groth).[24]

Subcontracting is commonly carried out locally, Inditex has more than 500 sewing contractors stationed close to its key factories, most of which work entirely for one of its brands. The working relationship is exceptionally close and Inditex scrutinises operations for quality, legal compliance, and timing before all sewn items are delivered to the factory for inspection, automated ironing, labeling, and wrapping prior to distribution.

Its market entry strategy is crafted with precision, foreign partners are chosen carefully

Even though its first joint venture in Italy with the Benetton Group did not work out, Inditex persevered to form suitable alliances with other companies. For instance, in 1998, Inditex created an alliance for the German and North European markets with the German company, Otto Versand. The strategy turned out to be successful. By 1999, the first German Zara store was open and three more were commissioned next year.

In more recent years, in 2008, Inditex created a 80-20 joint venture with the Lotte Group, in Korea, one of Asia's prime retailers. This venture will assist Zara to expand its business in Korea due to Lotte's experience of Korean market and opportunities of new retail locations (Zara opens its first stores in South Korea)[29].

In 2009, Inditex and the Tata Group signed an agreement to establish a joint venture to expand Zara stores into India.

Inditex controls just over half of the venture, and Tata Group owns the remaining 49 percent. Plans to start launching new stores in the major cities of India were developed for the year 2010. Tata Group claims that it sees excellent opportunities for Zara in a country which is becoming progressively more fashion aware. In addition, it states "Zara has proven itself capable of quickly adapting to changing consumer tastes across geographies. This association with brand Zara is our endeavor to bring a world class brand and experience to fashion conscious Indians" (Inditex and Trent of the Tata Group.[25]

Moreover, in 2009, Inditex and UNI Global Union signed a global agreement to officially distinguish their partnership, to guarantee workers' rights, and the growth and sustainability of the company. The contract between the pair is an approval of a range of fundamental rights of workers protected in ILO Conventions. Inditex and UNI claim their mutual interest in the growth and sustainability of the company, they acknowledge internationally standard conventions on human and trade union rights and they consent to work together to defend those rights (Inditex and UNI Global Union)[26].

To summarise, Inditex has modularised its outsourced production into small workshop units and sewing cooperatives that specialise in what they do best. Inditex is their only (or the biggest) customer in most cases and they respond extremely well to flex up and down with the demand pull flowing through the Inditex supply chain.

On the customer facing side, Inditex collaborates with

chosen leaders in facilitating local market entry in many countries where it makes sense. To these relationships, Inditex brings its unique blend of strategic imperative so that the overall business model of the 5-STAR network works uniformly across the entire business. Inditex would not hesitate to terminate a relationship if it does not work in a results focused manner - for example the failed relationship with the Benetton joint venture in Italy.

It is instructive to learn from the Inditex story. Each company will draw its own lessons based on its unique situation. However here are some common lessons that I think apply across the board for most businesses.

Ortega and the Inditex story has a number of critical lessons for CEOs, executives and entrepreneurs

1. Create your own business model. Never copy the business network or model of your competitors, suppliers or customers. Authenticity, originality, fit-for-purpose business models win the day every time.

2. Keep fine tuning your business model and the 5-STAR network continually as the circumstances change. Customer preferences evolve, newer suppliers become available, newer technologies come to the market and competitors change their strategies. All of these will require changes to your business model and the 5-STAR network.

3. Whenever there is a big change in your business - merger, acquisition, divestment, expansion, new market entry - re-consider the appropriateness of your business model and the 5-STAR network. What was appropriate until yesterday may no longer be appropriate.

4. Know your customers. The more you know your customer, the faster you can translate that knowledge into products they are willing and able to pay for, and the higher your ability to win in the marketplace. No matter how big your company becomes, do not lose touch with your customers. This customer knowledge is the primary source of innovation - keep it in front of everything your company does.

5. Make your suppliers dependent on your business. Become one of their biggest and best customers if you can. This way you can create a multitude of smaller suppliers - all part of your 5-STAR networks who will play the game as per your business strategy.

6. Use technology to its best effect. Keep investing in technology to gain maximum advantage towards your business strategy.

7. Create a buzz in the marketplace around your products, around your brand, and around your offerings. Keep the marketplace buzzing as much as possible via customer networks and conversations.

8. Outsource strategically, and modularise before outsourcing. Do not outsource to any single entity. Make outsourcing work for your business, now and in future.

There are many other lessons embedded throughout the long saga above, and each company and reader will take its own conclusions based on their own situation. The purpose of researching, collating and recounting this long case story is exactly to enable you to draw such lessons. I encourage you to write to me how you have utilised the information in this case study (or in the rest of this book) to make positive change in your business. This positive change could be minor tweaking of business network or relationship here and there; or, could be a major transformation exercise that continues over several months or years culminating in a thorough overhaul of the business model. Both are equally valid responses depending on the situation.

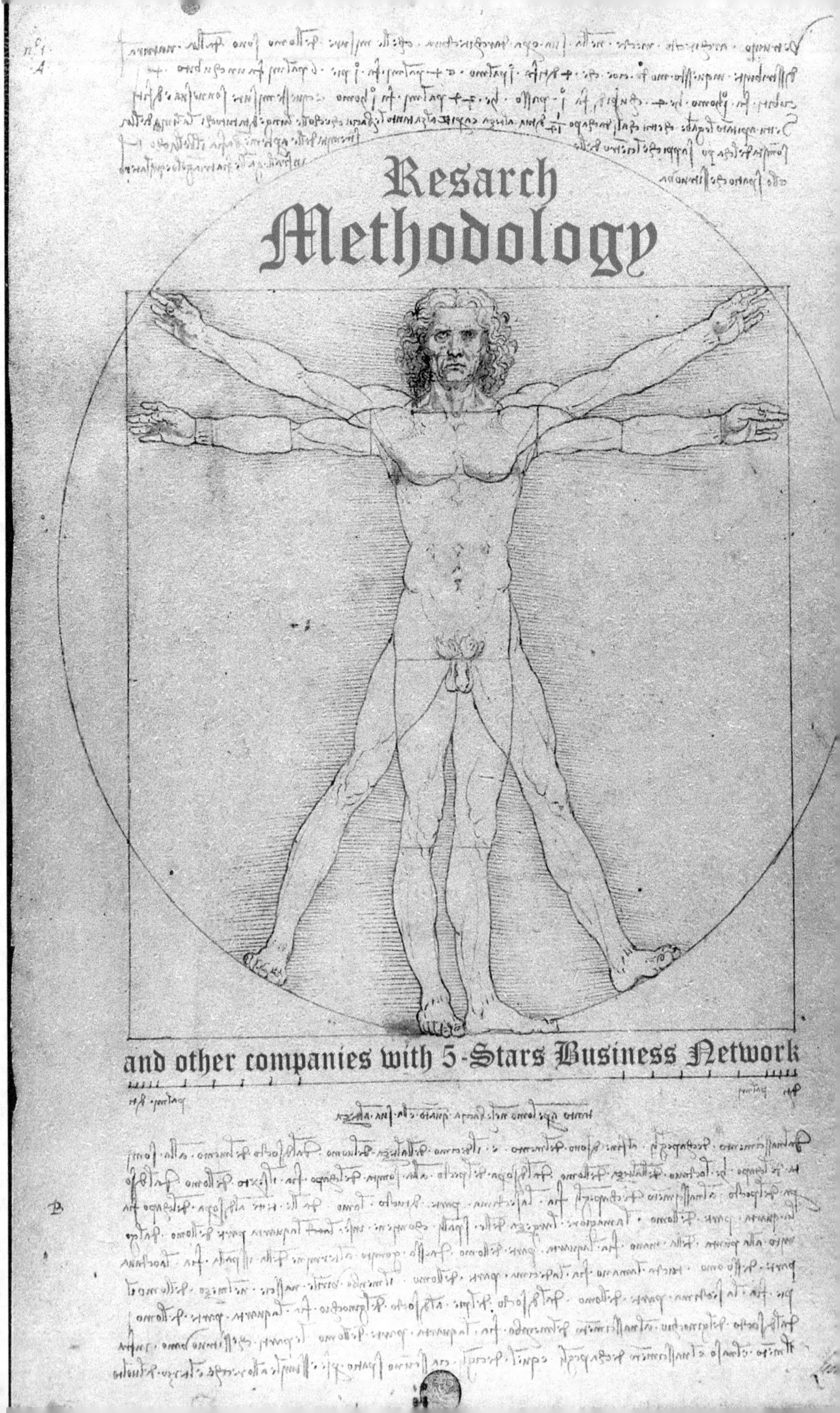

Resarch Methodology

and other companies with 5-Stars Business Network

Chapter 14

Research Methodology and Other Companies With 5-STAR Business Networks

Frequently the readers of books such as this come up with an objection about the companies used as case examples for the concepts. This is an understandable objection. What if your industry or geography is not represented in the case examples used through this book?

Does that mean that the concepts in the book are not useful for your company or situation?

Chances are that there are other companies - both in your industry and in your region - that can be used as a benchmark and example. For this reason, let me show you how our research team conducted the research to find suitable companies to look at for our work for this book. This is not a Warren Buffet style fundamental analysis for finding suitable investments.

To start with a caveat, this research is not focused on investigating suitable investment opportunities and should not be used as such. The sole focus of this research was to gauge the 5-STAR Business Networks of the companies on the five critical variables of interest to us: FAR Innovation, t Efficiency, TOP, APP and ROM. No databases are available with data for these variables, we used proxies for the purpose.

We started with a universe of nearly 1,200 companies and tracked their performance over five years - from 2007 to 2012. The reason for doing this is simple; most of your strategic decisions today will yield results only in three to five years time. Moreover, some of the variables, such as APP, measure the performance over a period of time covering several product life cycles.

Over the period of five years nearly, 20 percent of the companies dropped out of database - due to mergers, acquisitions, change of ownership, dropping out of top 1,200, etc. The rest of the database insurance companies and finance companies work in a highly regulated market environments and as such we found their data to be useless for comparison purposes. Without an inside peek into the internal workings of these companies, financial proxies would simply not be able to approximate for the five key variables of interest.

For the rest of the companies we used profit margins as the proxy for t Efficiency, five-year profit margin growth as the proxy for TOP, inverse of PE ratio (ratio of market value to profits, or market price as a multiple of profits) as a proxy for FAR Innovation, growth in inverse of PE ratio as a proxy for APP, and change in annual ranking over five years as a proxy for ROM.

Why did we use these proxies?

None of the proxies would have perfect correlation with the variables. In fact if we had to measure the variables in a single company we will have to come up with a composite of multiple internal variables based on internal processes and their outcomes. For this reason conducting a correlation test is not possible either.

Consider Figure 14.1 below which shows profits margins on sales for 2012 for 932 companies ranked in a descending order. The names of the companies are not shown on the figure, and all we are interested in is to divide this universe of companies into five sub-universes ranked by profit margins. The highest ranked 20 percent of the companies get a score of 5 on t Efficiency because indeed profit margins are derived from efficiencies.

Even if in some cases such as Amazon's margin of 1 percent, where the profit margins are very low placing them lower on t Efficiency measure than would be warranted from other internal variables, this deficiency will be made up in other measures such as market value to profits ratio (for

Figure 14.1: Profit margin 2012 for 932 companies

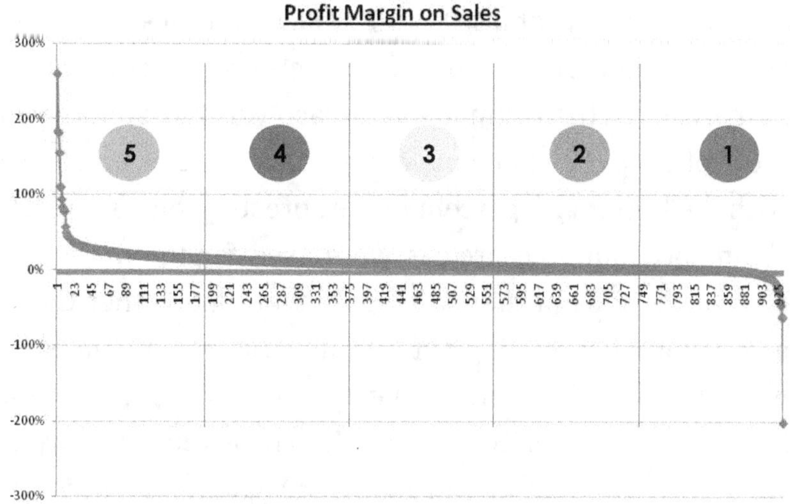

example in Amazon's case it is 125 percent compared to the market average of 14.9 percent).

Let us see how that happens.

Consider Figure 14.2 on the next page. This shows similar detail for market value to profits ratio for the same 932 companies in a descending order. Obviously the order has changed this time and Amazon is not middle of the pack in this figure; rather it is one of the top performers. We use this measure as a proxy for FAR Innovation - a variable which is notoriously very hard to define and measure. However, as a proxy, market price has built into it the expectations of future earnings coming from investment into useful and profitable innovation.

The proxy may not be perfect. For example, Amazon is ranked one of the highest on this measure, though there are other companies which might be far more innovative. However, when measured over a database of more than 1,000 companies and using rankings rather than absolute numbers for these variables, it is possible to obtain a discernable spread of data on each of the variables of interest. The top 20 percent of the companies for each variable were given a score of 5, next 20 percent were given a score of 4 and so on, with the bottom 20 percent of the spread given a score of 1.

Similarly suitable proxies were used for the other three variables - none of them are perfect, but altogether they do a good job of spreading out our universe of companies into meaningful sub-universes that you use for your purposes. After ranking each company on each of the five variable, we added the score and produced the company ranking list

Figure 14.2: Ratio of market value to profit in 2012 for 932 companies

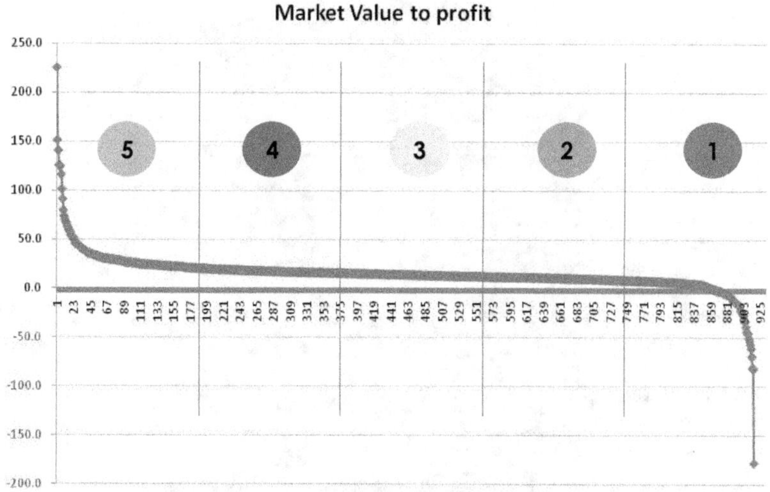

shown in Appendix 1.

So how do we use this table?

Consider Figure 14.3 on the next page.

The first thing that strikes us is that the sample incorporates companies from several different industries and different locations/regions. This is fairly typical of the entire universe of the companies researched.

This does raise some interesting questions about the companies shown (for example about future product development capacity of Apple), but we will not get into those questions in this book.

Our purpose for this analysis is to draw helpful hints. To start with - we notice there are companies from all industries and all over the world in this table. So readers who complain that they have no comparable companies can look at the closest match to their own operations and examine their operations in a lot

Figure 14.3: Sample out of top 100 companies with 5-STAR Business Network

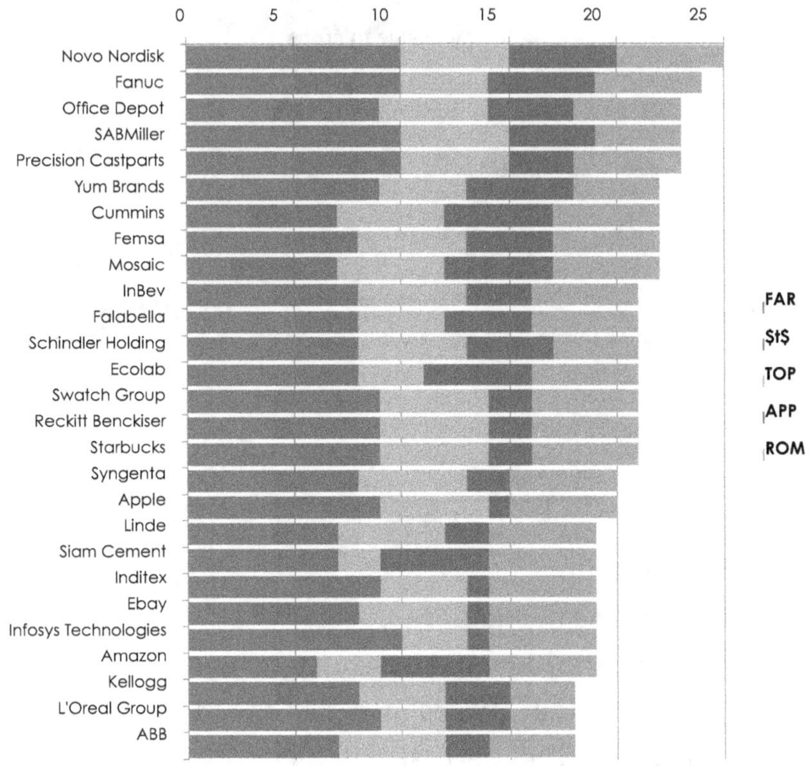

more detail to find discernable lessons for themselves.

Moreover, the key questions from the Inditex case studies are still relevant here. Ask the following questions:

1. How can I create my own business model without copying the business network or model of the comparable company that ranks highest in my industry or region? What can I learn from their business model and apply with authenticity and originality to create a fit-for-purpose business model for my business?

2. How can I fine tune my business model and the 5-STAR Business Network continually as the circumstances change? What have other companies in my industry and region done as customer preferences evolve, newer suppliers become available, newer technologies come to the market and competitors change their strategies?

3. How can I make sure that, whenever there is a big change in my business - merger, acquisition, divestment, expansion, new market entry, that my business model and the 5-STAR Business Network are re-created to correspond with the massive change?

4. How do I make sure that our people know our customers intimately? The more we know about our customers, the faster we can translate that knowledge into products customers are willing and able to pay for, the higher our ability to win in the marketplace.

5. How can we make the customer knowledge the primary source of innovation - keeping it in front of everything our company does?

6. How can we make our suppliers dependent on our business - becoming one of their biggest and best customers? What have others done to create a multitude of smaller suppliers as part of their 5-STAR Business Networks who play the game as per the masters' business strategy?

7. What are others doing in my industry or region to use

technology to its best effect? How can we keep investing in technology to gain maximum advantage towards our business strategy?

8. How can we create a buzz in the marketplace around our products, around our brand, and around our offerings? How do others keep the marketplace buzzing as much as possible via customer networks and conversations and social media?

9. How do others outsource strategically and modularise before outsourcing? How can we make outsourcing work for our business, now and in future?

There is nothing more powerful on earth than asking the right questions. If you choose the right companies from the table in the Appendix, research their operations well enough and think deeply on each of the questions given above, you will get a multitude of insights that will apply to your business. In case you need help doing so, please get in touch with me. Good luck.

APPENDIX

APPENDIX - ANALYSIS

Table of Companies with 5-STAR Business Networks

No.	Company	Country	Industry	t	FAR	ROM	APP	TOP	Total
1	Novo Nordisk	Denmark	Drugs & Biotechnology	5	5	5	5	5	25
2	Fanuc	Japan	Capital Goods	5	5	5	5	4	24
3	Office Depot	United States	Retailing	5	4	5	4	5	23
4	SABMiller	United Kingdom	Food Drink & Tobacco	5	5	4	4	5	23
5	Precision Castparts	United States	Aerospace & Defense	5	5	5	3	5	23
6	Union Pacific	United States	Transportation	5	4	4	4	5	22
7	Anheuser-Busch Cos	United States	Food Drink & Tobacco	4	4	5	4	5	22
8	National Oilwell Varco	United States	Oil & Gas Operations	4	4	5	4	5	22
9	EMC	United States	Technology Hardware & Equip	4	5	5	4	4	22
10	Yum Brands	United States	Hotels, Restaurants & Leisure	4	5	4	5	4	22
11	Goodrich	United States	Aerospace & Defense	4	4	5	5	4	22
12	NIKE	United States	Household & Personal Products	4	5	4	5	4	22
13	Pepsi Bottling Group	United States	Food Drink & Tobacco	4	4	5	4	5	22
14	British Amer Tobacco	United Kingdom	Food Drink & Tobacco	5	5	3	5	4	22
15	Cummins	United States	Capital Goods	4	3	5	5	5	22
16	Femsa	Mexico	Food Drink & Tobacco	3	5	5	4	5	22
17	Mosaic	United States	Chemicals	5	2	5	5	5	22
18	Vodafone	United Kingdom	Telecommunications Services	5	2	5	5	5	22

No.	Company	Country	Industry	t	FAR	ROM	APP	TOP	Total
19	Praxair	United States	Chemicals	4	4	4	4	5	21
20	AutoZone	United States	Retailing	4	4	5	4	4	21
21	Bed Bath & Beyond	United States	Retailing	4	4	5	4	4	21
22	VF	United States	Household & Personal Products	4	4	5	4	4	21
23	Occidental Petroleum	United States	Oil & Gas Operations	5	3	4	4	5	21
24	SAP	Germany	Software & Services	5	4	4	3	5	21
25	McDonald's	United States	Hotels, Restaurants & Leisure	5	4	3	4	5	21
26	InBev	Belgium	Food Drink & Tobacco	4	4	5	3	5	21
27	Falabella	Chile	Retailing	3	5	5	4	4	21
28	Schindler Holding	Switzerland	Capital Goods	3	5	4	4	5	21
29	TJX Cos	United States	Retailing	3	4	5	4	5	21
30	Grupo Mexico	Mexico	Materials	5	3	5	5	3	21
31	Ecolab	United States	Chemicals	3	5	5	5	3	21
32	Swatch Group	Switzerland	Household & Personal Products	5	4	5	2	5	21
33	Reckitt Benckiser	United Kingdom	Household & Personal Products	5	4	5	2	5	21
34	Starbucks	United States	Hotels, Restaurants & Leisure	4	5	5	2	5	21
35	Yamato Holdings	Japan	Transportation	2	4	5	5	5	21
36	Canadian National	Canada	Transportation	5	3	4	4	4	20
37	CSX	United States	Transportation	5	3	4	4	4	20
38	H&M Hennes & Mauritz	Sweden	Retailing	4	5	4	4	3	20
39	Danaher	United States	Conglomerates	4	4	5	3	4	20
40	Air Prods & Chems	United States	Chemicals	4	4	4	3	5	20
41	Black & Decker	United States	Household & Personal Products	3	4	5	5	3	20
42	Oracle	United States	Software & Services	5	4	4	2	5	20
43	Formosa Plastics	Taiwan	Chemicals	5	2	5	4	4	20
44	BG Group	United Kingdom	Oil & Gas Operations	5	4	4	5	2	20
45	Reynolds American	United States	Food Drink & Tobacco	5	4	2	4	5	20
46	Tatneft	Russia	Oil & Gas Operations	4	2	5	4	5	20

No.	Company	Country	Industry	t	FAR	ROM	APP	TOP	Total
47	Syngenta	Switzerland	Chemicals	4	4	5	2	5	20
48	Hyundai Mobis	South Korea	Consumer Durables	4	2	5	4	5	20
49	Amcor	Australia	Materials	2	5	5	4	4	20
50	Qualcomm	United States	Technology Hardware & Equip	5	5	5	3	2	20
51	ASML Holding	Netherlands	Semiconductors	5	2	5	3	5	20
52	Coach	United States	Household & Personal Products	5	5	5	2	3	20
53	Delphi	United States	Consumer Durables	3	2	5	5	5	20
54	Boston Scientific	United States	Health Care Equipment & Svcs	3	5	2	5	5	20
55	Enbridge	Canada	Oil & Gas Operations	2	5	5	5	3	20
56	Nikon	Japan	Business Services & Supplies	2	5	5	5	3	20
57	Wharf (Holdings)	Hong Kong/ China	Conglomerates	5	1	5	4	5	20
58	Public Storage	United States	Diversified Financials	5	5	4	1	5	20
59	Goldcorp	Canada	Materials	5	4	5	1	5	20
60	Apple	United States	Technology Hardware & Equip	5	4	5	1	5	20
61	Monsanto	United States	Chemicals	4	5	5	1	5	20
62	EOG Resources	United States	Oil & Gas Operations	4	5	5	5	1	20
63	Agilent Tech-nologies	United States	Technology Hardware & Equip	4	4	4	3	4	19
64	Walt Disney	United States	Media	4	4	3	4	4	19
65	BCE	Canada	Telecommunications Services	4	3	4	4	4	19
66	Telus	Canada	Telecommunications Services	4	4	4	4	3	19
67	Aire Liquide Group	France	Chemicals	4	4	4	3	4	19
68	General Mills	United States	Food Drink & Tobacco	4	4	4	3	4	19
69	Henkel Group	Germany	Household & Personal Products	3	4	4	4	4	19
70	Norfolk Southern	United States	Transportation	5	3	3	4	4	19
71	IBM	United States	Software & Services	4	3	3	4	5	19
72	Hershey	United States	Food Drink & Tobacco	4	5	3	4	3	19
73	Dassault Aviation	France	Aerospace & Defense	3	5	3	4	4	19
74	Luxottica Group	Italy	Household & Personal Products	3	5	4	4	3	19
75	Nidec	Japan	Capital Goods	3	5	5	3	3	19

No.	Company	Country	Industry	t	FAR	ROM	APP	TOP	Total
76	Honeywell International	United States	Conglomerates	3	5	3	5	3	19
77	Sumitomo Trust & Banking	Japan	Banking	4	5	2	4	4	19
78	Colgate-Palmolive	United States	Household & Personal Products	4	4	4	2	5	19
79	Pernod Ricard	France	Food Drink & Tobacco	4	4	4	2	5	19
80	Schlumberger	Netherlands	Oil & Gas Operations	4	5	4	4	2	19
81	Express Scripts	United States	Health Care Equipment & Svcs	2	4	5	4	4	19
82	Cnooc	Hong Kong/ China	Oil & Gas Operations	5	3	5	4	2	19
83	Baxter International	United States	Health Care Equipment & Svcs	5	3	4	2	5	19
84	Grupo Modelo	Mexico	Food Drink & Tobacco	4	5	2	5	3	19
85	Canadian Pacific Railway	Canada	Transportation	4	5	3	5	2	19
86	Linde	Germany	Conglomerates	3	4	5	2	5	19
87	Technip	France	Oil & Gas Operations	3	4	5	2	5	19
88	Siam Cement	Thailand	Construction	3	4	5	5	2	19
89	Japan Tobacco	Japan	Food Drink & Tobacco	3	5	4	5	2	19
90	China Resources Ent	Hong Kong/ China	Conglomerates	2	5	5	4	3	19
91	Barrick Gold	Canada	Materials	5	2	5	2	5	19
92	Keppel	Singapore	Conglomerates	5	2	5	2	5	19
93	Wm Morrison Supermarkets	United Kingdom	Food Markets	2	2	5	5	5	19
94	Delta Air Lines	United States	Transportation	2	2	5	5	5	19
95	China Unicom	Hong Kong/ China	Telecommunications Services	2	5	5	5	2	19
96	Woodside Petroleum	Australia	Oil & Gas Operations	5	4	5	4	1	19
97	Spectra Energy	United States	Oil & Gas Operations	5	4	4	1	5	19
98	Atlas Copco	Sweden	Capital Goods	5	4	4	5	1	19
99	Richemont	Switzerland	Household & Personal Products	5	5	4	4	1	19
100	Stryker	United States	Health Care Equipment & Svcs	5	4	5	1	4	19
101	Teva Pharmaceutical Inds	Israel	Drugs & Biotechnology	4	4	5	1	5	19
102	Ford Motor	United States	Consumer Durables	4	1	5	4	5	19

No.	Company	Country	Industry	t	FAR	ROM	APP	TOP	Total
103	Inditex	Spain	Household & Personal Products	4	5	5	1	4	19
104	Broadcom	United States	Semiconductors	4	5	5	1	4	19
105	Thermo Fisher	United States	Health Care Equipment & Svcs	4	4	5	1	5	19
106	Potash of Saskat-chewan	Canada	Chemicals	5	3	5	1	5	19
107	Infineon Tech-nologies	Germany	Semiconductors	5	1	3	5	5	19
108	Ebay	United States	Retailing	5	3	5	1	5	19
109	Infosys Tech-nologies	India	Software & Services	5	5	5	1	3	19
110	General Motors	United States	Consumer Durables	3	1	5	5	5	19
111	Tyson Foods	United States	Food Drink & Tobacco	1	3	5	5	5	19
112	Amazon	United States	Retailing	1	5	5	5	3	19
113	Publicis Groupe	France	Media	4	3	4	3	4	18
114	Emerson Electric	United States	Conglomerates	4	4	3	3	4	18
115	Hutchison Whampoa	Hong Kong/ China	Conglomerates	4	4	3	3	4	18
116	SKF Group	Sweden	Capital Goods	4	3	3	4	4	18
117	Kellogg	United States	Food Drink & Tobacco	4	4	3	3	4	18
118	Eaton	United States	Capital Goods	3	3	4	4	4	18
119	Parker-Hannifin	United States	Capital Goods	3	3	4	4	4	18
120	Schneider Electric	France	Capital Goods	3	4	4	4	3	18
121	Bayer Group	Germany	Chemicals	3	4	4	3	4	18
122	Cooper Industries	Bermuda	Capital Goods	4	3	3	3	5	18
123	Mattel	United States	Household & Personal Products	4	3	3	3	5	18
124	L'Oréal Group	France	Household & Personal Products	4	5	3	3	3	18
125	Dover	United States	Conglomerates	4	3	3	3	5	18
126	Celanese	United States	Chemicals	3	3	3	4	5	18
127	Franklin Resources	United States	Diversified Financials	5	3	4	2	4	18
128	Canadian Natural Res	Canada	Oil & Gas Operations	5	3	4	4	2	18
129	Novartis	Switzerland	Drugs & Biotechnology	5	4	3	4	2	18
130	St Jude Medical	United States	Health Care Equipment & Svcs	4	4	5	2	3	18

No.	Company	Country	Industry	t	FAR	ROM	APP	TOP	Total
131	Inpex Holdings	Japan	Oil & Gas Operations	4	4	5	3	2	18
132	Royal DSM	Netherlands	Chemicals	4	2	3	4	5	18
133	Unilever	Netherlands/ United Kingdom	Food Drink & Tobacco	4	4	3	5	2	18
134	Scana	United States	Utilities	4	3	2	4	5	18
135	ABB	Switzerland	Capital Goods	3	4	4	2	5	18
136	Accenture	Bermuda	Business Services & Supplies	3	4	4	2	5	18
137	PPR	France	Retailing	3	4	4	2	5	18
138	Christian Dior	France	Household & Personal Products	3	4	5	2	4	18
139	Lyondell Chemical	United States	Chemicals	2	3	5	4	4	18
140	CVS	United States	Retailing	2	4	5	3	4	18
141	Sodexho Alliance	France	Business Services & Supplies	2	5	4	3	4	18
142	SES Global	Luxembourg	Media	5	3	2	3	5	18
143	Amgen	United States	Drugs & Biotechnology	5	3	3	2	5	18
144	Roche Holding	Switzerland	Drugs & Biotechnology	5	3	3	2	5	18
145	MTN Group	South Africa	Telecommunications Services	5	3	5	3	2	18
146	Sanofi-aventis	France	Drugs & Biotechnology	5	3	3	2	5	18
147	Carlsberg	Denmark	Food Drink & Tobacco	3	3	5	2	5	18
148	Sandvik	Sweden	Capital Goods	3	5	3	5	2	18
149	Apache	United States	Oil & Gas Operations	5	2	5	4	2	18
150	Marathon Oil	United States	Oil & Gas Operations	5	2	2	4	5	18
151	Xstrata	Switzerland	Materials	5	2	4	2	5	18
152	Automatic Data	United States	Business Services & Supplies	4	5	2	5	2	18
153	Kühne & Nagel Intl	Switzerland	Transportation	2	5	5	2	4	18
154	Konica Minolta	Japan	Business Services & Supplies	2	4	2	5	5	18
155	Galp Energia SGPS	Portugal	Oil & Gas Operations	2	5	4	5	2	18
156	Abbott Laboratories	United States	Drugs & Biotechnology	4	4	4	1	5	18
157	CLP Holdings	Hong Kong/ China	Utilities	4	4	4	5	1	18
158	Hyundai Motor	South Korea	Consumer Durables	4	1	4	4	5	18

No.	Company	Country	Industry	t	FAR	ROM	APP	TOP	Total
159	Rio Tinto	United Kingdom/ Australia	Materials	4	4	4	5	1	18
160	Costco Wholesale	United States	Retailing	1	5	4	4	4	18
161	Adobe Systems	United States	Software & Services	5	5	4	1	3	18
162	Diageo	United Kingdom	Food Drink & Tobacco	5	4	3	5	1	18
163	Bristol-Myers Squibb	United States	Drugs & Biotechnology	5	3	4	1	5	18
164	Texas Instruments	United States	Semiconductors	5	4	3	5	1	18
165	Yara International	Norway	Chemicals	4	1	5	3	5	18
166	Mitsui Sumitomo Ins	Japan	Insurance	4	5	1	3	5	18
167	Noble Energy	United States	Oil & Gas Operations	4	5	3	5	1	18
168	Rogers Communications	Canada	Telecommunications Services	4	3	5	1	5	18
169	Fidelity National Info	United States	Business Services & Supplies	3	4	5	1	5	18
170	Formosa Chems & Fibre	Taiwan	Chemicals	3	4	5	5	1	18
171	Merck	Germany	Drugs & Biotechnology	3	5	4	5	1	18
172	Swisscom	Switzerland	Telecommunications Services	3	5	4	5	1	18
173	Google	United States	Software & Services	5	5	5	1	2	18
174	Las Vegas Sands	United States	Hotels, Restaurants & Leisure	5	5	5	1	2	18
175	Safran	France	Aerospace & Defense	2	5	5	1	5	18
176	VimpelCom	Russia	Telecommunications Services	2	5	5	5	1	18
177	Mitsubishi Motors	Japan	Consumer Durables	1	5	2	5	5	18
178	3M	United States	Conglomerates	4	3	3	4	3	17
179	Groupe Danone	France	Food Drink & Tobacco	3	4	4	3	3	17
180	HJ Heinz	United States	Food Drink & Tobacco	3	4	3	3	4	17
181	United Technologies	United States	Conglomerates	3	4	3	3	4	17
182	Deere & Co	United States	Capital Goods	3	3	4	3	4	17
183	Caterpillar	United States	Capital Goods	3	3	4	4	3	17
184	PPG Industries	United States	Chemicals	3	3	4	3	4	17
185	United Parcel Service	United States	Transportation	3	4	3	4	3	17
186	Kubota	Japan	Capital Goods	3	4	3	4	3	17

No.	Company	Country	Industry	t	FAR	ROM	APP	TOP	Total
187	Johnson & Johnson	United States	Drugs & Biotechnology	4	4	3	4	2	17
188	Merck & Co	United States	Drugs & Biotechnology	4	4	4	3	2	17
189	Nan Ya Plastic	Taiwan	Chemicals	4	3	4	4	2	17
190	PepsiCo	United States	Food Drink & Tobacco	4	4	4	3	2	17
191	DirecTV Group	United States	Media	4	3	4	2	4	17
192	Shin-Etsu Chemical	Japan	Chemicals	4	4	4	2	3	17
193	Komatsu	Japan	Capital Goods	3	4	4	2	4	17
194	Limited Brands	United States	Retailing	3	4	2	4	4	17
195	Genuine Parts	United States	Consumer Durables	2	4	3	4	4	17
196	Isuzu Motors	Japan	Consumer Durables	2	4	3	4	4	17
197	Chunghwa Telecom	Taiwan	Telecommunications Services	5	3	3	4	2	17
198	Medtronic	United States	Health Care Equipment & Svcs	5	3	4	2	3	17
199	Anglo American	United Kingdom	Materials	5	2	3	3	4	17

If your company, suppliers, customers, or partners are not in the table above, they may still be in the database - please contact the author for the information you are seeking...

NOTES

CHAPTER 1

1. Kim, Lilian. "Price of Prius jumps in wake of Japan disasters." *Abclocal.go.com.* 18 March, 2011. Web. **http://abclocal.go.com/kgo/story?section=news/business&id=8020150**
2. Lanahan, Lawrence. "The Disaster Tab." *Carey.jhu.edu.* 4, 1. 2011. Web. **http://carey.jhu.edu/one/2011/fall/the-disaster-tab/**
3. "Financial and Inventory Strategies for Working Capital Optimisation." *Aberdeen.com*. Oktober, 2007. Web. **http://aberdeen.com/_aberdeen/Working percent20Capital percent20Optimisation/-/-/-/search.aspx**
4. Thomas, Mark. "Fortescue Refinances for Rapid Growth." *Fmgl.com.* 10 Oct., 2010. Web. **http://www.fmgl.com.au/Investors_and_Media/media_releases/News_Details/Fortescue_Refinances_for_Rapid_Growth_144**
5. Safe Bulkers. "The international drybulk shipping industry is cyclical and volatile..." *Wikinvest.com.* 8 June, 2009. Web. **http://www.wikinvest.com/stock/Safe_Bulkers_ percent28SB percent29/International_Drybulk_Shipping_Industry_Cyclical_Volatile_Charter_Rates**

6. "Marketing Essays: Red Bull." *Ukessays.com. 2013.* Web. **http://www.ukessays.com/essays/marketing/red-bull.php**

CHAPTER 2

1. Major, John. "Trade along the Silk Roads: The Silk Route-2000 Miles of History." New York. 1996. Print.
2. Puga, Diego & Trefler, Daniel. "International Trade and Institutional Change: Medieval Venice's Response to Globalisation." *Economic History and International Trade and Regional Economics.* August, 2012. Web.
3. Stein, Stanley J. "Bureaucracy and Business in the Spanish Empire, 1759-1804: Failure of a Bourbon Reform in Mexico and Peru." *The Hispanic American Historical Review.*" 61,1. Feb, 1981. Web.
4. Innes, AD. "The East India Company: A History of the British Nation." *Britainexpress.com.* 1912. Web. http://**www.britainexpress.com/History/East-India-Company.htm**
5. "Transcontinental Railroad: Building Change." *Schmoop.com.* nd. Web. **http://www.shmoop.com/transcontinental-railroad/summary.html**

CHAPTER 3

1. Gladwell, Malcolm. *The Tipping Point: How Little Things can make a Big Difference.* New York. Back Bay Books. 2002. Print.
2. Dixit, Avinash & Nalebuff, Barry. *Thinking Strategically: The Competitive Edge in Business, Politics, and Everyday*

Life. New York. W.W Norton & Company. 1993. Print.

3. "Network Effect." *Investitopia.com.* 2013. Web. **http://www.investopedia.com/terms/n/network-effect.as**

CHAPTER 4

1. Sood, Vivek. "A Comprehensive Report on Strategic Trends in Supply Chain Information Technology." CreateSpace 2010. Print.

CHAPTER 5

1. Kawaja, Terry. "Social Media Marketing." *Businessinsider.com.* 2012. Web. **http://www.google.com.au/imgres?um=1&hl=en&sa=N&tbo=d&biw=1600&bih=809&tbm=isch&tbnid=bbgv_qStqWtOgM:&imgrefurl=http://www.businessinsider.com/social-media-marketing-landscape-complicated-2012-5&docid=Ey75dEOvtUgdvM&imgurl=http://static1.businessinsider.com/image/4fb5077becad045f47000003-960/buddy-media-social-marketing.jpg&w=960&h=708&ei=D9caUe_ZMuWViAeL5oCoAg&zoom=1**

CHAPTER 6

1. Hagel III, John., & Singer, Marc. "Unbundling the Corporation." *The McKinsey Quarterly.* Volume 3. (2000). 148-161. Web. 4 Feb. 2013.

CHAPTER 7

1. "Global Trends." *National Intelligence Council.* 2012. Web. **http://www.dni.gov/index.php/about/organisation/national-intelligence-council-global-trends**
2. Anders, George. "Jeff Bezos's Top 10 Leadership Lessons." *Forbes.com.* 4 April, 2012. Web. **http://www.forbes.com/sites/georgeanders/2012/04/04/bezos-tips/**
3. "Bezos on Innovation." *Bloomberg Businessweek Magazine.* 16 April, 2008. Web. **http://www.businessweek.com/stories/2008-04-16/bezos-on-innovation**
4. Lashinsky, Adam. "Amazon's Jeff Bezos: The Ultimate Disrupter." *Fortune Magazine.* 16 Nov, 2012. Web. **http://management.fortune.cnn.com/2012/11/16/jeff-bezos-amazon/**
5. "Amazon: The Hidden Empire." *faberNovel: Ideas with legs.* 2013. Web. **http://www.slideshare.net/faberNovel/amazoncom-the-hidden-empire**
6. Copeland, Lee. "Borders Turns to Amazon for Outsourcing." *Computerworld.com.* 16 April, 2001. Web. **http://www.computerworld.com/s/article/59638/Borders_Turns_to_Amazon_for_Outsourcing**
7. Fitzgerald, Michael. "Amazon Finds Profits in Outsourcing." *CIO.com.* 15 Oct. 2002. Web. **http://www.cio.com/article/31447/Amazon_Finds_Profits_in_Outsourcing**

CHAPTER 8

1. Christensen, Clayton. *The Innovator's Dilemma*. New York. HarperBusiness. 2011. Print.
2. Christensen, Clayton. *The Innovator's Solution.* New York. HarperBusiness. 2011. Print.
3. Christensen, Clayton. *The Innovator's Prescription.* New York. HarperBusiness. 2011. Print.
4. Mulder, Karel. *Sustainable Development for Engineers.* Austin, Texas. Greenleaf Publishing. 2006. Print.
5. Krishan, M.S., Prahalad, C.K. *The New Age of Innovation: Driving Co-Created Value through Global Networks.* New York. McGraw Hill. 2008. Print.

CHAPTER 9

1. Dell, Michael, Fredman, Catherine. *Direct from Dell: Strategies That Revolutionized an Industry.* New York. HarperBusiness. 2006.Print.

CHAPTER 10

1. Hill, Kashmir. "How Target Figured Out A Teen Girl Was Pregnant Before Her Father Did." *Forbes.com.* 16 Feb, 2012. Web. **http://www.forbes.com/sites/kashmirhill/2012/02/16/how-target-figured-out-a-teen-girl-was-pregnant-before-her-father-did/**
2. Wilson, Timothy. "Can't Help Myself: 'The Power of Habit.'" *Nytimes.com*. 9 March, 2012. Web. http://www.nytimes.com/2012/03/11/books/review/the-power-of-habit-by-charles-duhigg.html?pagewanted=all&_r=0\

3. Duhigg, Charles. *The Power of Habit: Why We Do What We Do in Life and Business.* [Toronto]: Doubleday Canada, 2012. Print.

CHAPTER 11

1. Reuters. "Intel's strategy to stay on top of rivals. *Times of India.* 19 Jan. 2013. Web. **http://articles.timesofindia.indiatimes.com/2013-01-19/strategy/36431857_1_intel-tablet-and-phone-markets-stacy-rasgon**
2. "Intel Tick Tock Strategy- What it Means." *Podtech.net.* 12 Sep. 2008. Web. **http://www.podtech.net/home/5356/intel-tick-tock-strategy-what-it-means**
3. "Intel Corporation: The Evolution of an Adaptive Organisation." *Aomonline.org.* nd. Web. **http://meeting.aomonline.org/1999/INTEL1.htm**
4. Burkitt, Laurie. "Inside Intel's Marketing Strategy." *Forbes.com.* 18 May 2010. Web. **http://www.forbes.com/2010/05/18/intel-marketing-creators-project-vice-cmo-network-deborah-conrad.html**
5. Hensley, Muthusami, et al. "Transforming Intel's Supply Chain to Meet Market Challenges." *Intel.com.* Jan. 2012. Web. **http://www.intel.com.au/content/dam/www/public/us/en/documents/best-practices/transforming-supply-chain-to-meet-market-challenges.pdf**

CHAPTER 12

1. "Transocean Ltd. Reports Fire on Semisubmersible Drilling Rig Deepwater Horizon." *Deepwater.com.*

Transocean, Inc. 21 April, 2010. **http://www.deepwater.com/fw/main/Transocean-Ltd-Reports-Fire-on-Semisubmersible-Drilling-Rig-Deepwater-Horizon-448C936.html?LayoutID=46**

2. Repanich, Jeremy. "The Deepwater Horizon Spill by the Numbers." *Popular Mechanics.* Hearst Communication, Inc. 10 August, 2010. **http://www.popularmechanics.com/science/energy/coal-oil-gas/bp-oil-spill-statistics**
3. Boulden, Jim. *BP still wants Halliburton to pay for whole Gulf oil spill cleanup. CNN News.* Turner Broadcasting System, Inc. 5 Jan., 2012. **http://edition.cnn.com/2012/01/03/us/bp-halliburton-lawsuit/index.html**
4. CNN Wire Staff. "BP ready to try again to stop oil flow, company says." *CNN News.* Turner Broadcasting System, Inc. 15 May, 2010. **http://edition.cnn.com/2010/US/05/14/gulf.oil.spill/index.html**
5. McMahon, Paul Anthony. "Southern Cross University Candidate's Declaration." *Tendering for Outsourced Services in a Large Australian Bank: The Drivers of Success.* Southern Cross University. Sep. 2008. Web. **http://www.byronconsulting.com.au/PaulMcMahon_SCU_DBA_Thesis.pdf**

CHAPTER 13

1. Walt,Vivienne. *Fortune Magazine.* 8 Jan. 2013. Print.
2. Erikkson, C., & Jonsson, S. (2011). *Inditex: A company analysis with focus on growth.* Retrieved

June 18, 2012, from **http://gupea.ub.gu.se/bitstream/2077/25661/1/gupea_2077_25661_1.pdf**

3. "Business Lessons from Highly Successful Entrepreneurs - Amancio Ortega." *Small Business Database*. 27 Feb. 2012. **http://www.smartbusinessdatabase.com/entrepreneurial-tips/business-lessons-from-highly-successful-entrepreneurs-amancio-ortega/**
4. "Growth Through Innovation – The Inditex Experience." *GO Boldly*. **http://www.retail.org.nz/downloads/Inditex.pdf**
5. Craig, A., Jones, C., Nieto, M. "Zara: Fashion Follower, Industry Leader." *Philadelphia University*. 2 April 2004. **http://2011.fashion-networks.com/images/article_pdf/ZaraReport.pdf**
6. Sloane, Paul. "Agile innovation lessons from Zara." *Innovation Tool.com*. 2012. **http://www.innovationtools.com/Articles/EnterpriseDetails.asp?a=763**
7. Kihlen, Tobias. "On Logistics in the Strategy of the Firm." *UniTryck, Linköping*. 2005. **http://www.google.com/url?sa=t&rct=j&q=&esrc=s&source=web&cd=8&ved=0CF4QFjAH&url=http percent3A percent2F percent2Fliu.diva-portal.org percent2Fsmash percent2Fget percent2Fdiva2 percent3A20334 percent2FFULLTEXT01&ei=TevfT6iRDbC4iAf69cWICg&usg=AFQjCNF9IZ8QS0Fe2Ht5pKBCVRf490KHQg&sig2=yHPaR95GSh0r9nvgVroKww**
8. Katigbak, N., Labog, R., Leung, K., Li, R., Ortiz, M., Papilla, M. "Zara: Fast Fashion Case Study." *Scribd.com*. 2011. **http://www.scribd.com/doc/79423996/Case-Study-**

Zara-Fast-Fashion

9. "EMPRESAS: Inditex, The Fashionable Company. "*Plantilla Simple.* 26 Sept 2011. **http://mundodelaempresa.blogspot.com.au/2011/09/empresas-inditex-fashionable-company-in.html**
10. Hitt, M., Ireland, R., and Hoskisson, R. "Strategic Management: Competitiveness and Globalisation: Concepts & Cases." *South Western.* 8th Edition. 2009. **http://books.google.com.au/books?id=ul5FsIlWa3EC&pg=PA118&lpg=PA118&dq=inditex+-+cost+leadership&source=bl&ots=2V29EAzTPw&sig=m6gCwabVtftH7fioZtqagwvan3I&hl=en&sa=X&ei=8njiT8y_Bs3HmQWSr93NAw&ved=0CFEQ6AEwAA#v=onepage&q=INDITEX&f=false**
11. Dopico, D., and Porral, C. "Analysis of Value Chain and Sources of Differentiation in International Fashion Markets." *European Research Studies.* 2011. **http://www.google.com.au/url?sa=t&rct=j&q=&esrc=s&source=web&cd=4&ved=0CFkQFjAD&url=http percent3A percent2F percent2Fwww.gcd.udc.es percent2Fcristina percent2Fsubido percent2Fpublicaciones percent2F73 percent2F20120104120639_ Value percent2520Chain percent2520in percent2520Fashion percent2520Markets.doc&ei=8H3iT56FI8j7mAWajtH4Aw&usg=AFQjCNFWz76c3Y62D5z1DZ99tMt9-CoBzA&sig2=MvmOJwvppSSJFSkWPIrn1w**
12. Pearson, A. "The Story of Zara: The Speeding Bullet." *Unique Business Strategies.* Pages 1-2. **http://www.**

uniquebusinessstrategies.co.uk/pdfs/case percent20studies/zarathespeedingbullet.pdf

13. Jannoun, R. "Zara: Case Study." *Scribd.com.* **http://www.scribd.com/doc/55583327/Zara-Final-Search**
14. "Inditex." **http://www.mchodge.com/documents/Operations percent20Management/Zara.**
15. Caro, F., and Gallien, J. "Zara Uses Operations Research to Reengineer Its Global Distribution Process." August 2009. **http://web.mit.edu/jgallien/www/ZaraInterfacesPaperAugust3.pdf**
16. Shehzade, N. "Zara: International Business Analysis." **http://www.scribd.com/doc/15561231/Zara-spanish-apparel-brand**
17. "Inditex (ITX: Continuous)." *BloombergBusinessWeek.* 30 April 2012. **http://investing.businessweek.com/research/stocks/financials/ratios.asp?ticker=ITX:SM**
18. "Brand Watch: Zara." *Devangsbu Dutta.* 12 February 2004. **http://www.thirdeyesight.in/articles/Deconstructing_Zara_handout_version.pdf**
19. "Management for Marketers." *The Marketing Institute.* 15 August 2008. **http://www.mii.ie/attachments/wysiwyg/36/E08AugCIMS-4.pdf**
20. "Zara Production." *OtherPapers.com.* 2011. **http://otherpapers.com/Business/Zara-Production/29863.html**
21. Mukherjee, N., Sujit, N., Janmeja, S., Mohata, R., Rani, S., Shinde, P. "Supply Chain Practices of Zara." *Scribd.com.* 2010. **http://www.scribd.com/doc/19488507/Supply-Chain-Practices-at-Zara**

22. Mukherjee, N., Sujit, N., Janmeja, S., Mohata, R., Rani, S., and Shinde, P. "Supply Chain Practices of Zara." *Scribd.com*. 31 August 2009. **http://www.scribd.com/doc/19488507/Supply-Chain-Practices-at-Zara**

23. Choudhury, B., and Holmgren, T. "Supply Chain Agility towards Global Outsourcing within Fast Fashion industry." *Jönköping*. May 2011. **http://scholar.google.com.au/scholar?q=outsourcing**

24. Goransson, S., Jonsson, A. and Perrson, M. "Extreme Business-Models in the Clothing Industry." *Kristianstad University*. December 2007. **http://www.google.com.au/url?sa=t&rct=j&q=&esrc=s&source=web&cd=10&ved=0CF8QFjAJ&url=http percent3A percent2F percent2Fhkr.diva-portal.org percent2Fsmash percent2Fget percent2Fdiva2 percent3A230949 percent2FFULLTEXT01&ei=MfToT8foBoiEmQX8xsiJDg&usg=AFQjCNHPWBFuJgoSOYLyEyVnYgXnJtDecA&sig2=U0kOPa4OcUVsjlX8SE272Q**

25. Groth, A. "This Apparel Brand Realised the Big Disadvantage of Outsourcing To China." *Business Insider, Inc.* March 2012. **http://articles.businessinsider.com/2012-03-27/strategy/31243435_1_pablo-isla-supply-chain-retailer-zara**

26. "Inditex and Trent of the Tata Group agree to open stores in India beginning 2010." *Inditex Annual Report 2011*. 2 May 2009.

27. "Inditex and UNI Global Union sign global deal to guarantee workers' rights." *InditexAnnual Report 2011*. 2 October 2009. **http://www.inditex.com/en/press/**

press_releases/extend/00000751

28. Webinar, G. "New Supply Chain Strategies to Maximise Product Value and Growth- Gartner Webinar Review- Part 1." *Lasta.* 9 December 2011. **http://www.esourcingforum.com/archives/2011/12/09/new-supply-chain-strategies-to-maximise-product-value-and-growth-gartner-webinar-review- percentE2 percent80 percent93-part-1/**
29. "Inditex: A strenuous supply chain." *Taha Zia.* May 2012. **http://economy247.blogspot.com.au/2012/05/inditex-one-of-strongest-supply-chains.html**
30. "Inditex Group." *El Rincón del Vago.*1998. **http://html.rincondelvago.com/inditex-group.html**
31. "Zara opens its first stores in South Korea." *Inditex Annual Report 2011.* 29 April 2008. **http://www.inditex.com/en/press/press_releases/extend/00000615**

CHAPTER 14

1. Maurer, Bill. "Skeptical of Amazon at these levels." *Seekingalpha.com.* 1 May, 2012. **http://seekingalpha.com/article/544591-skeptical-of-amazon-at-these-levels.pdf+and+inditex&hl=en&as_sdt=0&as_vis=1&oi=scholart&sa=X&ei=gOvoT9mSKYXdmAXVor2mDg&ved=0CFIQgQMwAA**

ACKNOWLEDGEMENT

This book is the distillation of the experiences and knowledge I have gained from working, reading, applying concepts, learning, experimenting and observing results across multiple decades. All this could not have happened in isolation and depended on many incredible people. Many people who I have worked with, who have written books that I read, who participated in numerous workshops, training and coaching sessions and who otherwise collaborated with me in business interactions have left some mark on this book. If you fall under one of these categories, you know who you are - and I gratefully acknowledge your contribution to the evolution of my thinking reflected in this book.

Today, as I write this acknowledgement on the day celebrated as the Teachers' Day across India in memory of one of our great teachers, I am reminded of all the great teachers I have had.

First and foremost of these is the Greatest Teacher of the Universe, who sends the right lesson packaged as an intriguing problem at the right moment in life. I count myself as one of the luckiest people on earth - and these lessons are the biggest reason I do so.

My next teacher in life was my mother whose persistence and relentless thinking was exemplary and infectious. In a part of the world where little girls are still gunned down for daring to ask for education, she managed to do the near impossible. In an era when girls were sternly told that learning to read and write was the end of education rather than the start, she fought insurmountable odds to complete a high school diploma, a bachelor's degree, a master's degree in economics and then another master's degree in sociology - and then she made it her mission to provide knowledge and opportunities to girls across the country.

My father taught me the lesson of self-reliance so well that I easily left home to start work at the age of 18 and went on to self-fund all my higher education and training. Even after his untimely death, his encouragement, trust and counsel continue to give me the self-confidence and assurance to make the toughest decisions today without any hesitation. If I can teach my three boys even half of what he taught me, I would consider my debt repaid.

My wife Mamta has been a constant source of compassion and joy who miraculously manages to create time for me to travel, work, lecture, conduct workshops, write and study for up to 18 hours a day for several weeks in a row. All this amid a career of her own and educating our three young boys cheerfully. I cannot acknowledge enough her contribution to all my activities, including writing this book.

I was fortunate to learn from the masters - both in the education institutions I attended, and in the workplace. To paraphrase Sir Isaac Newton - ***If I have seen far, it is because I was standing on the shoulders of the giants.*** It is impossible to name all of these giants but I will try to name a few in no particular order: Dr. Wolfgang Partsch, Gene Tyndall, Ben Gordon, Garry Baddock, Malcolm Broomhead, Shaun O'Sullivan, Julian Segal, Philippe Etienne, Neil Hawker, Hugh Robertson, John D'Souza, Ian Alexander, Marc Potter, Parag Mehta, Hanns Zeltinger, Martyn Hay, Gavin Solsky, Jeremy Barrett, Daniel Wise, Cluny Randall, Andrew Liveris, Essa Al Saleh, Samir Wagdy, Mark Aquilina, Andrew Wilson, Jim Camp, Frank Cascante, Paul Lim, Stephen Chan, Alex Bauer, Stephen Chey, Nick Bain, Bernhard Voll, Kevin Daly, Norm Fricker and many others - contributed to insights and ideas that are crystallized throughout this book.

In addition, during the formative years, Mr. Aggarwal (our Chemistry teacher in year 7 introduced us to the writings of Ayn Rand and changed the direction of my career in that year), Alistair MacLean (the writer whose novels gave me a strong yearning for sea by the time I was in year 10), Captain Rewari (whose humorous lectures on navigation and ship stability made the nautical college tolerable), Fred Hilmer (the Dean of Australian Graduate School of Management who somehow created and ran an institution far above any other in the country at that time), Simon Wheately (the lecturer for International Finance in MBA for his bravery in persisting with the concepts when half the class was getting overwhelmed), Tim Jackson and Wayne Henderson at Booz Allen and Hamilton for showing me the ropes in the new world of management consulting,

Dr. Wolfgang Partsch and many other colleagues who have collaborated with me on projects and workshops around the world, and finally the clients and participants in the workshops who have always shared their enthusiasm and thoughts on the topics under discussion.

This book started as an attempt to answer to a few simple questions that we were brainstorming in a small team with David Ball and Estelle Berger. Some of the answers grew into the concepts you see throughout this book. Estelle Berger then took on the additional responsibility of fleshing out the ideas into a PowerPoint presentation describing the core story woven through the pages of this book. As the idea of writing this book based on the PowerPoint came to me, I invited Lalit Panda, a key industry expert to co-write this book. His initial commitment, could not translate into actual writing given his busy schedule and we will likely co-write a sequel, on the implementation of the concepts expounded in this book.

Tony Fedorowicz has kindly edited parts of the first edition of this book - if you find it easy to understand what I write, it is mainly due to his effort to simplify my writing. Kelsey Claflin and James Barger have helped with, research and analysis for parts of this book. Nick Georgandis and Geoff Adams did the final edits and proofing for a professional production. Many other experts have provided feedback, suggestions for improvement and testimonials - which are all gratefully received and acknowledged. Finally, Jhernilyn Velasco, Tayyaba Satti and their teams have created the artwork, the print ready proof and the e-book edition ready for publication.

It has a humungous effort to write, produce and publish a book and put it into your hands. I gratefully acknowledge the help I got all along the way and hope I can return the favor in an equal manner.

INDEX

t 102, 135, 138, 139, 141, 163, 179, 182, 183, 189, 192, 193, 195, 212, 290, 293, 316, 317

$eed-to-$tore 100, 144, 178, 181, 183, 184, 185, 189, 190, 288, 290

5-STAR Business Networks IT systems 3

5-STAR network 17, 55, 157, 268, 269, 293, 294, 306, 310, 311

A

Adam Smith 54, 55

Aerospace 15

Airbus A380 46

Alfred Chandler 54, 57

Alphonse KARR 21

American railroads 2, 27, 30

Apple XXI, 16, 37, 55, 56, 111, 114, 115, 116, 122, 148, 149, 150, 151, 173, 181, 215, 229, 319

architecture 4, 88, 90, 91, 92, 94, 96, 110

Asia XXV, 22, 25, 126, 279, 290, 307, 308
Automotive 15, 48
Ayn Rand 273, 347

B

Barbalet 45, 49
Barings 28
Blow-out Prevention 240
BP 238, 240, 241, 339
budgeting process 78
business cycles 29, 75
business network XX, 2, 8, 9, 10, 12, 13, 15, 16, 22, 23, 24, 25, 26, 27, 29, 30, 37, 70, 75, 83, 94, 101, 102, 104, 122, 173, 174, 178, 188, 195, 210, 211, 212, 225, 254, 262, 268, 273, 292, 310, 312, 320
business networks XI, XV, XXII, XXX, XXXIII, XXXVIII, XXXIX, 1, 2, 3, 4, 8, 9, 10, 12, 13, 14, 15, 16, 17, 18, 21, 22, 28, 29, 30, 33, 42, 45, 51, 59, 60, 66, 82, 83, 93, 95, 96, 99, 100, 102, 105, 123, 128, 130, 134, 137, 175, 181, 204, 211, 212, 213, 252, 254, 257, 267
Business Process Outsourcing 244

C

capital XXIX, 13, 16, 179, 180, 227, 230, 231, 275, 290, 291, 297, 298
Carl Jung 52
Charles Mingus 65
Chicago Times-Herald 48
China XXIII, XXXV, XXXVI, 22, 26,

28, 253, 293, 307
Chinese 22, 28, 150, 255
C.K. Prahalad 57
Columbus 25
competitive advantage XXXVI, 2, 18, 82, 111, 112, 180, 193, 220, 222, 229, 232, 234, 256, 282, 284, 290, 293, 295
competitors 8, 71, 77, 79, 102, 119, 130, 139, 148, 150, 152, 179, 181, 195, 228, 231, 233, 251, 268, 284, 286, 287, 289, 296, 297, 298, 303, 304, 307, 310, 321
complexity 3, 67, 68, 95, 105, 134, 249, 257, 264
complexity and risks 69
cooperation 27, 31, 104
cost of service 203, 208, 245
cost conscious 77
customer preferences 268, 310, 321
customer satisfaction 86
customer service 120, 142, 189, 208, 244

D

David Ball 72, 348
Deepwater Horizon 237, 238, 239, 240
demand 23, 25, 68, 69, 75, 76, 84, 86, 87, 154, 160, 161, 165, 167, 168, 169, 170, 171, 172, 173, 189, 192, 193, 194, 209, 219, 223, 246, 257, 275, 285, 287, 291, 297, 301, 302, 303, 309
Dietrich Mateschitz 15, 16, 61
different XXIV, XXV, 36, 40, 55,

67, 72, 75, 87, 112, 128, 153, 166, 168, 197, 205, 208, 230, 251, 257, 262, 281, 282, 285, 286, 319

E

East India Company 2, 26, 27, 30

economic principles 56

electronics 9, 15, 111, 133, 136, 216, 295

Ericsson 9, 10, 12, 148

ERP systems 204

Europe XXXIV, 23, 24, 26, 250, 274, 284, 289, 290, 292, 293, 297, 306, 307

F

Facebook XXIX, XXXIX, 15, 34, 45, 46, 51

fashion businesses 274

fax machines 49

finance XXIII, XXIV, 2, 13, 24, 28, 30, 245, 316, 347

FAR Innovation 102, 135, 151, 172, 174, 175, 178, 211, 215, 299, 316, 318

Fire-Aim-Ready 100, 137, 144, 148, 149, 150, 164, 168, 173, 283

forecasting 75, 76, 84, 87, 90, 192, 302

Fortescue Metals 13

free markets 3, 55, 58, 59

G

General Motors XXII, XXIII, 7

Gladwell 47

global network 174, 274

H

homogeneous groups 46

I

Inditex 173, 225, 274, 275, 276, 277, 278, 279, 280, 281, 282, 283, 284, 285, 287, 288, 289, 290, 291, 292, 293, 294, 295, 296, 297, 298, 299, 300, 301, 302, 303, 304, 305, 306, 307, 308, 309, 310

information technology 4, 66, 67, 69, 70, 72, 73, 75, 76, 77, 83, 94, 95, 96, 103, 185, 213

IT 3, 4, 33, 65, 66, 67, 70, 71, 72, 73, 75, 76, 77, 79, 81, 82, 83, 92, 95, 96, 129, 141, 149, 199, 244, 260

innovation XXXVII, 86, 101, 102, 104, 111, 116, 118, 119, 120, 121, 130, 131, 134, 135, 136, 137, 139, 140, 141, 149, 150, 151, 152, 153, 156, 157, 164, 165, 166, 167, 168, 170, 171, 172, 173, 174, 175, 179, 181, 185, 215, 216, 217, 218, 219, 220, 222, 224, 225, 235, 269, 283, 285, 286, 287, 288, 299, 311, 318, 321

iPhone 12, 93, 149, 150, 151, 172, 173, 283

J

Japan XXIV, 9, 84

Japanese economy 9

joint venturing 243

JP Morgan 28

K

Keynes 54, 56

Kiddy's Class 277

Korea 308

L

Leonardo da Vinci 267

linkedin XXIX, XXXIX, 38, 40, 51, 52

logistics XXV, XXIX, 13, 62, 86, 87, 113, 141, 174, 189, 194, 212, 244, 245, 249, 259, 282, 283, 284

M

Malcolm Gladwell 46

Managed Markets 58, 59

Managed Networks 59

market segmentation 279, 298

Mark Zuckerberg 15

Mediterranean Sea 24

Michael Dell 15, 177

middle ages 2, 22, 23, 24, 25, 29

mining XXXVI, 13, 15, 27

mistakes XV, 3, 4, 70, 72, 73, 78, 95

modularisation 101, 105, 141, 144, 252, 254, 255, 257, 258, 259, 262, 264

MRP revolution 84, 86

N

network effect 3, 45, 46, 47, 48, 49, 50

network IT systems 3, 66, 72

Networks XXII, XXXVIII, 2, 3, 8, 21, 24, 30, 46, 51, 52, 58, 88

Nokia XXI, 9, 10, 12, 45, 111, 148

North America XXII, 27, 48, 290

O

oil 28, 198, 237, 238

Oracle 136

Ortega 225, 273, 275, 283, 299,

310

outsourcing 105, 141, 142, 238, 242, 252, 312

P

PC industry 102, 179, 195, 257

Peter Drucker XXII, XXXVIII

Philips 9, 10

plug and play 91

Programming 70, 71

proprietary systems 92

Pull & Bear 278, 279

Pullman 27

PUSH model 84

R

Ralph Waldo Emerson 267

Results-based Outsourcing 251

Results-focused Outsourcing 252, 259, 261

ROM 105, 261

retail 138, 144, 184, 223, 283

Rothschild 28

S

Service Level Agreements 250

silk road 2, 22

smartphone 282

social networks XXXVIII, 2, 34, 42, 283

software 4, 7, 67, 70, 88, 94, 232

Spanish 275, 276, 283, 287

Spanish empire 2, 25, 29

Stanford 27, 47

Steve Jobs 15, 55, 56, 147

Stradivarius 281

Stuart Emmett 52

super networked business 17, 100, 121, 123, 144, 185, 193

supply chain management XIX, 91, 287, 305

supply chains XXV, 66, 83, 104, 161, 218, 222, 289
supply network management 4, 88
Supply Network Management Strategy 4, 83, 96
supply networks 2, 36, 42, 70, 87, 95, 121, 224, 268
synchronicity 3, 52, 57, 59, 61
synchronization 87, 91, 92
synergy 3, 49, 51, 53, 55, 59

T

Tata Group 308, 309
Technology 99, 118, 311
time and cost 80
Traditional economy 86
TOP 93, 103, 135, 140, 179, 210, 212, 316
Transaction Optimisation Profitability 101, 144, 193, 294

U

utility 42, 49, 50

V

Value accretion 3, 53
value synchronicity 3, 53
value synergy 3, 53
Vanderbilt 27
Vasco da Gama 25
Venetian 24, 25
Venetian republic 24
Venice 2, 22, 24, 25, 29

Z

Zara 275, 277, 278, 279, 282, 283, 284, 285, 286, 297, 306
Zara Kids 277

WANT MORE?

Be Part of the Conversation

Log on to www.5starbusinessnetwork.com
Send us an email: info@5starbusinessnetwork.com

Or Clip this page - tick a box, fill in your contact details and fax it, take a photo with your smartphone and email it, or send an MMS/SMS:

YES, I want more information about how 5-STAR Business Networks can help my business.
Please contact me about:

- ☐ 5-STAR Business Networks
- ☐ Innovation: Fire-Aim-Ready (FAR) Innovation
- ☐ Cash Actualisation: $eed-to-$tore (t) Cycle Management
- ☐ Beneficial Growth: Transaction Optimisation Profitability (TOP)
- ☐ Product Serialisation: Advanced Product Phasing (APP)
- ☐ Insourced Outsourcing: Results-focused Outsourcing and Modularisation (ROM)
- ☐ Other ____________________

Name ____________________

Title ____________________

Company ____________________

Website ____________________

Email ____________________

Telephone ____________________

Other comments:

www.ingramcontent.com/pod-product-compliance
Lightning Source LLC
LaVergne TN
LVHW052222200926
848481LV00001B/184
* 9 7 8 0 6 1 5 7 9 4 1 9 8 *